Manual of Water Supply Practices

M21

Groundwater Operations

Fifth Edition

American Water Works Association

M21 — Groundwater Operations, Fifth Edition

Groundwater Operations

Managing Editor – Book Products: Melissa Valentine
Manager – Manuals: Molly Beach
Manager – Publishing Operations: Gillian Wink
Specialist – Copyright and Permissions: Peggy Tyler
Director of Publishing: John Fedor
Technical Editor: Wiley
Cover Composition: Michael Labruyere
Cover Art: Feaspb/Shutterstock.com
Production: Innodata

Library of Congress Cataloging-in-Publication Data
Names: American Water Works Association, editor.
Title: M21 - groundwater operations / by American Water Works Association.
Other titles: Ground water (American Water Works Association) | Groundwater operations
Description: Fifth edition. | Denver, CO : American Water Works Association, [2025] | Includes bibliographical references and index.
Identifiers: LCCN 2023053087 (print) | LCCN 2023053088 (ebook) | ISBN 9781647171490 (paperback) | ISBN 9781613006788 (adobe pdf)
Subjects: LCSH: Wells. | Groundwater.
Classification: LCC TD405 .M25 2025 (print) | LCC TD405 (ebook) | DDC 628.1/14--dc23/eng/20231220
LC record available at https://lccn.loc.gov/2023053087
LC ebook record available at https://lccn.loc.gov/2023053088

Printed in the United States of America

ISBN 978-1-64717-149-0 ISBN, electronic 978-1-61300-678-8

https://doi.org/10.12999/AWWA.M21ed5

American Water Works Association

American Water Works Association
6666 West Quincy Avenue
Denver, CO 80235-3098
303.794.7711
awwa.org

Contents

List of Figures

List of Tables

This page intentionally blank.

Preface

More than 41,000 water systems rely on wells, many of them small systems. The fifth edition of M21 is the culmination of nearly five years of effort by members of the Groundwater Resource Committee to focus the manual's contents on water utility operational applications. AWWA has published other manuals and books that cover topics previously found in M21, such as a mathematical summary of groundwater, a book on well drilling, a summary of wellfield modeling, and a treatise on aquifer storage or other tools. The goal of this revised manual is to provide the reader with a general understanding of the operation of wells; the monitoring of groundwater as a means to identify problems; a summary of the principles involved with groundwater, its movement, and character; and the subsequent effect these characteristics have on the well systems for water utilities. The contents of this edition are substantially different than previous editions, and there are several reasons for the changes.

Groundwater protection, planning, and evaluation efforts have evolved in the past 10 years as has modeling. As a result, much of this discussion of modeling has been shifted to the recently produced M73. Likewise, in 2016, AWWA and the committee produced M63 on aquifer storage and recovery so that material was also removed from M21. Members of the committee developed a book, *Water Supply Wells: Siting, Drilling and Construction,** which covers various issues with siting and drilling water supply wells; as a result, much of that material was also removed from M21. The ANSI/AWWA A100 standard remains as a companion to this manual and the drilling book for those who need more information on drilling processes. The result is a more operations-focused, shorter guide for M21, retitled Groundwater Operations. This manual is intended to meet the needs of operations staff and engineers for the 41,000 groundwater systems, providing sufficient information on the operations and monitoring of their wells to help them with decision-making.

Scope

Chapter 1 is an overview of the occurrence and behavior of groundwater, including the geology, hydrologic cycle, and aquifer characteristics that define groundwater flow. A brief discussion is included on the process to evaluate aquifers and water quality.

Chapter 2 outlines the type and construction of wells that can be used for water supplies for utilities and describes the types of pumps used in well applications; maintenance requirements; pump problems; and solutions to those problems.

Chapter 3 discusses the recordkeeping used with wells and wellfield systems. These records provide utility personnel with insight into the occurrence of problems and long-term trends. This chapter has seen the most revision since previous editions.

Chapter 4 outlines the problems likely to be encountered as well as plugging and fouling problems and their correction. Microbiological fouling is a major topic

* Bloetscher, F., Muniz, A. and Largey, J. 2022. *Water Supply Wells: Siting, Drilling and Construction.* Denver: AWWA.

discussed in detail in this chapter as it is a major issue throughout the world—albeit one that is not commonly understood.

Chapter 5 summarizes water treatment issues arising from groundwater sources. The discussion is not exhaustive, but it describes common treatment options that operators, engineers, and administrators of water supply agencies should be aware of, may use, or may need to consider. It expands on the information of the previous versions of M21.

Chapter 6 summarizes the manual and where more information is available if more details are needed.

This manual should help operators and engineers gain enough background on the subject of groundwater to improve their decision-making. The manual should help these professionals answer many of their questions about complex aquifer systems and improve their response to problems. The Groundwater Resource Committee is hopeful that the new edition will meet the industry's needs and will be as useful as the previous editions have been.

Frederick Bloetscher, PhD, PE
Chair, AWWA Groundwater Resource Committee

Acknowledgments

The AWWA Technical & Educational Council, the Water Resource and Sustainability Division, and the Groundwater Resource Committee gratefully acknowledge the contributions made by those volunteers who drafted, edited, and provided the significant and critical commentary essential to updating M21. The M21 subcommittee members dedicated many hours in the preparation of this fifth edition to ensure the overall technical quality, consistency, and accuracy of the manual.

Contributors

Frederick Bloetscher (Chair), Florida Atlantic University, Boca Raton, Fla.
Angie Brown, AWWA, Denver, Colo.
David Edson (Vice Chair), Retired, Jensen Beach, Fla.
Dawn Flancher, AWWA, Denver, Colo.
Sharon Long, Retired, University of Wisconsin–Madison, Middleton, Wis.
Sandra M. Eberts, US Geological Survey, Columbus, Ohio
Albert Muniz, Hazen and Sawyer, Boca Raton, Fla.

This page intentionally blank.

Chapter 1

The Occurrence and Behavior of Groundwater

This chapter provides a general overview of:

- the hydrologic cycle;
- the general terminology used by engineers and hydrogeologists to describe aquifer systems;
- general groundwater concepts;
- the interrelationship of groundwater and various geologic formations with surface water, precipitation, and drainage patterns; and
- the effect of groundwater on water supplies.

Because groundwater and surface water resources are closely related, any event that occurs aboveground can impact an underground water supply. As a result, utilities that use groundwater supplies need to monitor surface events, such as rainfall, spills, accidents, changes in land use, development, fire, and drought to determine the potential effect on their water systems. Utilities have an important role in understanding, protecting, and responsibly managing groundwater resources in the face of changing climate and socioeconomic conditions.

HYDROLOGIC CYCLE

All living things depend on water for survival. Of the total water found on earth, 97.3 percent is saltwater in the oceans. Of the remaining water, more than two-thirds exists as ice in the polar caps currently, although average annual sea ice has begun to decline. The rest, or 0.61 percent of all water, is fresh water in lakes, rivers, streams, and groundwater.

The constant movement of water above, on, and below the earth's surface is defined as the hydrologic cycle and is depicted in Figure 1-1 (for more detailed diagrams of the hydrologic cycle, see the US Geological Survey website). From oceans and seas, to clouds, to precipitation, to streams and lakes, to groundwater, and back to the oceans, water travels through the hydrologic cycle, conditioning and shaping all that it touches. The hydrologic cycle is the conceptual model used in the development and management of water supplies. The components of the hydrologic cycle are:

- evaporation and transpiration,
- precipitation,
- surface water and runoff, and
- groundwater.

Evaporation and Transpiration

Although the hydrologic cycle is continuous and has neither a beginning nor an end, evaporation and transpiration will be discussed first in this manual. These two processes are commonly combined and referred to as the process of evapotranspiration (ET). ET is the process of water vapor entering the atmosphere through water that evaporates from open water bodies and through water that transpires from vegetation. ET rates vary, depending largely on the amount of solar radiation, the latitude of the catchment area, temperature, water surface area, and vegetative cover. Areas close to the equator tend to have higher ET rates as shown in Figure 1-2. In subtropical areas during the wet season or during summer months in northern latitudes, large bodies of water, including wetlands and estuarine areas, have high evaporation rates. This rising moisture forms clouds that condense and return the water to the land surface or oceans in the form of precipitation. Open water of shallow depth has the highest evaporation rate. Water as far as 4 ft below the surface may be subject to evaporation.

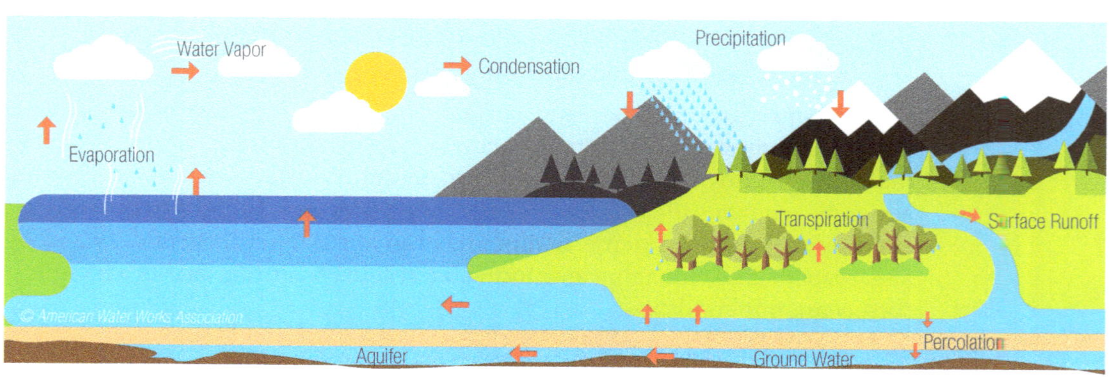

Source: DrinkTap.org by AWWA

Figure 1-1 Hydrologic cycle

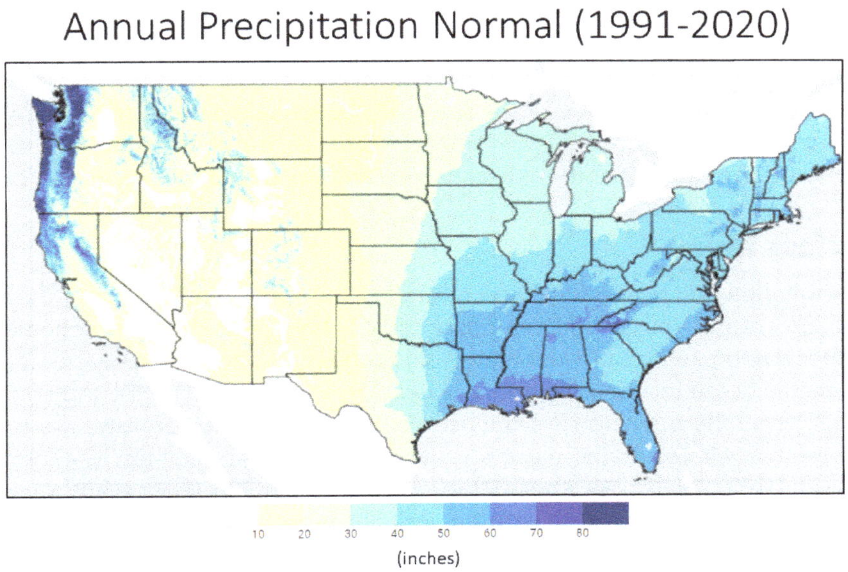

Source: NOAA

Figure 1-4(b) Annual precipitation normal (1991–2020)

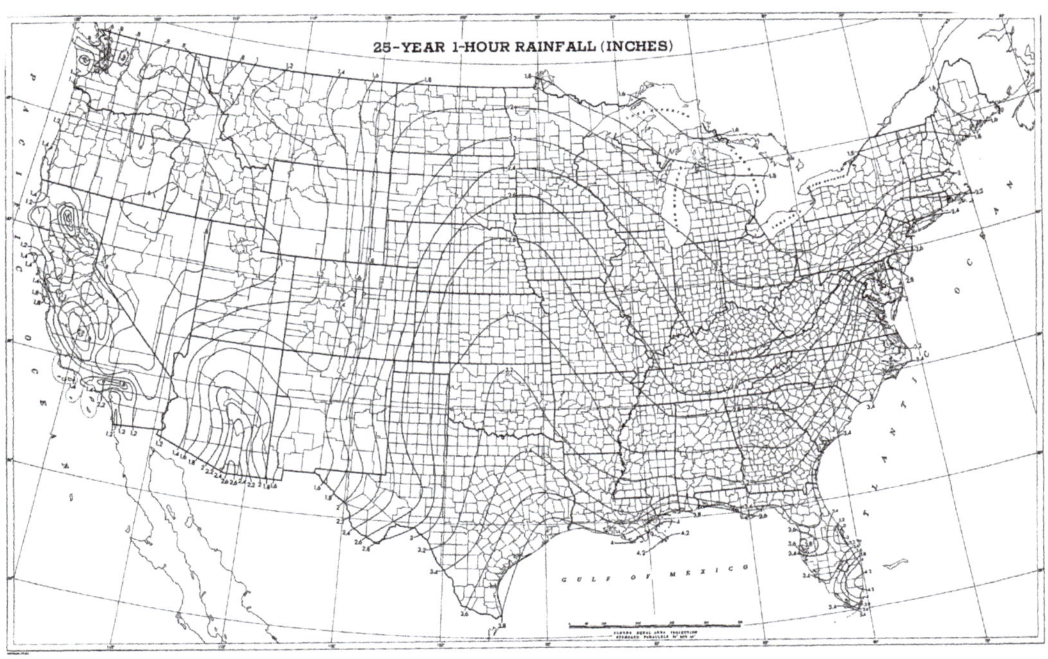

Source: NOAA 1963

Figure 1-5 One-hour precipitation events in the contiguous 48 states, 1935–1960

Those fronts depend on temperature differences to deliver precipitation. When it is hot and dry, little rainfall will occur because the water available for evaporation is limited. The western half of the United States is drier than the eastern. Figure 1-5 shows rainfall intensities that occur in the continental United States. Rainfall intensity is greatest in the Southeast. When the rate of precipitation exceeds the rate of infiltration, overland flow—or runoff—occurs.

Precipitation falls on the land surface where it wets vegetation and other surfaces, then it either runs off (to surface water bodies such as rivers and lakes) or infiltrates into the ground. Infiltration rates vary widely, depending on land use, development, the character and moisture content of the soil, and the intensity and duration of the precipitation event. Infiltration rates can vary from as much as 1 in./h (25 mm/h) in mature forests on sandy soils, to almost nothing in clay soils and areas that are built upon or paved.

Surface Water

Precipitation that runs off the land reaches streams, rivers, and lakes. If the water percolates into the ground, the groundwater may still discharge into these water bodies. Surface water bodies that mix with saltwater bodies along the coast are called estuaries (e.g., where a river meets the ocean in a delta). Surface water flow is controlled by the topography because the water generally flows toward the oceans. Just as precipitation rates are typically lower in the western United States, natural water bodies in this region are less common than in the eastern United States. As a result, the US government has spent many years and billions of dollars creating reservoirs to support agriculture west of the Mississippi River.

Infiltration and Recharge

Precipitation that percolates downward through porous surface soils is the primary source of water for groundwater. Surface areas having this downward flow are called "recharge areas." The characteristics of soil depend on the parent material, the climate, the types of organisms in and on the soil, the topography of the land, and the amount of time these factors have acted on the material. Because vegetative types differ in their nutrient requirements and in their ability to live in water-saturated or saline areas, soil types also play a role in determining plant-type distribution. Development or land uses play a particularly significant role in recharge—buildings and paved parking areas prevent infiltration from occurring, thereby increasing runoff and the potential for flooding.

Soil has the capacity to absorb some moisture initially, a factor called "initial infiltration." Initial infiltration replaces moisture in the root or plant zones, where the roots for most vegetation exist. Because of the variable permeability and transmissivity of different soils, the rate of groundwater recharge from precipitation will vary. For an aquifer to have fresh water, there must be a source of recharge, some degree of flow (albeit slow), and a discharge area (to cause the flow). Otherwise, if there were no recharge or movement, the aquifer would become brackish through dissolution of the minerals in the rock.

BEHAVIOR OF GROUNDWATER

Unsaturated and Saturated Zones

After precipitation has infiltrated the soil, it will travel down through two zones. The unsaturated zone, which occurs immediately below the land surface in most areas, contains water and air. The unsaturated zone is almost invariably underlain by a zone in which all interconnected openings in the ground are full of water. This zone is referred to as the saturated zone and is illustrated in Figure 1-6.

Water in the saturated zone, technically called "groundwater," is water that has the potential to surface in natural springs under certain geological conditions. It is usually insufficient in volume to be used as a community water supply. Recharge of the saturated zone occurs by percolation of water from the land surface through the unsaturated zone. Therefore, the unsaturated zone is of great importance to groundwater hydrology. This zone may be divided into three parts: the soil zone, the intermediate zone, and the upper part of the capillary fringe.

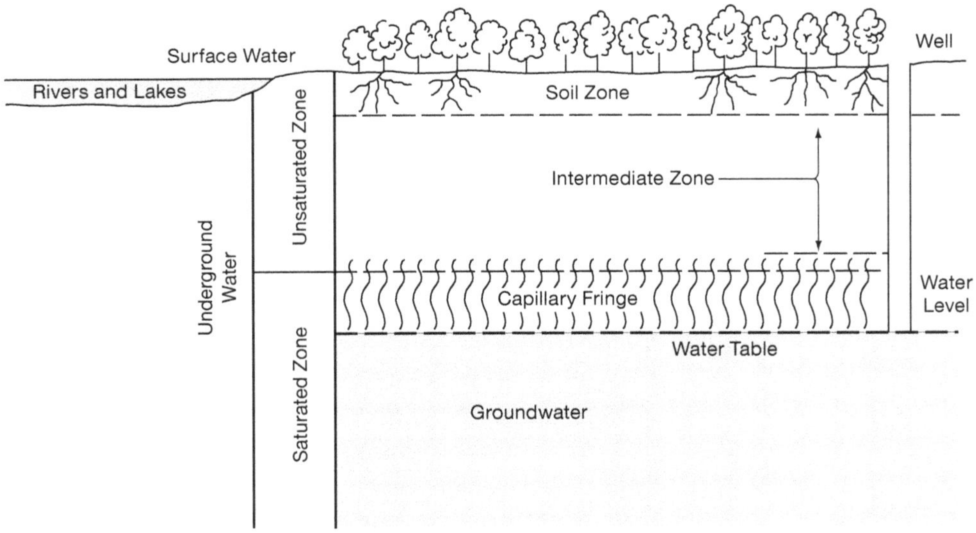

Figure 1-6 Water movement below the earth's surface

Soil zone. The soil zone typically extends from the land surface to a maximum depth of 3–5 ft (1–1.5 m). The soil zone supports plant growth; it is crisscrossed by living roots, voids left by decayed roots of earlier vegetation, and animal and worm burrows. The porosity and permeability of the material in this zone tend to be higher than the porosity and permeability of the ground beneath it.

Intermediate zone. Below the soil zone is the intermediate zone, which differs in thickness from place to place depending on the thickness of the soil zone and the depth to the capillary fringe. The intermediate zone is less porous than the soil zone since few roots or burrows penetrate it.

Capillary fringe. The capillary fringe is the subzone between the unsaturated and saturated zones. The capillary fringe occurs when a film of water clings to the surface of rock and soil particles and rises in small-diameter pores against the pull of gravity. Water in the capillary fringe and in the overlying part of the unsaturated zone is under a negative hydraulic pressure (i.e., less than atmospheric [barometric] pressure). The water table is the water level in the saturated zone at which the hydraulic pressure is equal to atmospheric pressure.

Aquifers and Confining Beds

Below the unsaturated soil zone, all solid materials ("rocks," including unconsolidated sediments) under the earth's surface can be classified either as aquifers or as confining beds (or confining units for geologists). An aquifer is rock (or formation) that will yield water in a usable quantity to a well or spring. Some of the groundwater pumped has been stored in aquifers for hundreds or thousands of years. A confining bed is a rock formation that restricts the movement of groundwater either into or out of adjacent rock formations as shown in Figure 1-7.

Groundwater occurs in aquifers under two different conditions. Near the land surface, water only partly fills the exposed formation. The upper surface of the saturated zone is free to rise and decline in direct relation to precipitation. The water in this type of an aquifer is unconfined, and the aquifer is an unconfined or water table aquifer. Wells that pump water out of unconfined aquifers are water-table wells. The water level in these

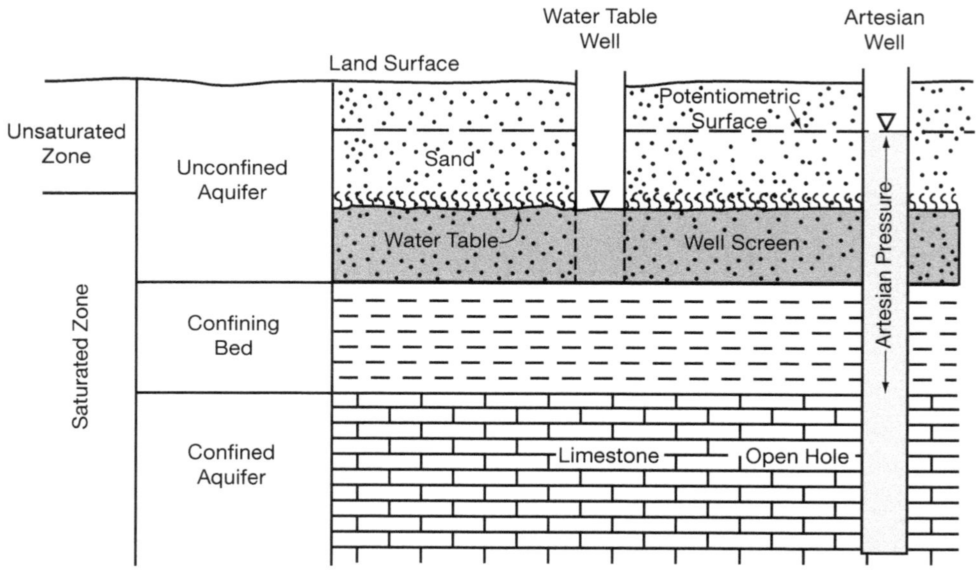

Figure 1-7 Geologic configuration of aquifers and confining beds

wells indicates the position of the water table in the surrounding aquifer. Wellfields that use water table aquifers can be recharged by rain and snowmelt. They are also vulnerable to contamination.

Under natural conditions, groundwater moves downhill until it reaches the land surface at a spring or through a seep along the side or bottom of a stream channel or estuary—or, in deeper aquifers, the ocean. Groundwater in the shallowest part of the saturated zone moves from interstream areas toward streams or the coast. In many areas, the direction of groundwater movement can be derived from observations of land topography when the land slopes toward water bodies. Thus, the water table typically replicates the land surface except where it intersects surface water bodies as shown in Figure 1-8.

Below the water table aquifer are confining units. Clay layers are a good example of a confining unit (generally an impervious formation), meaning water cannot flow through them easily. Clay consists of very small particles and electrochemically holds water molecules. As a result, clay is saturated with water but permits virtually no flow. Metamorphic material such as shale can also form a confining layer. The ability of water to flow through the soil is referred to as transmissivity. Clay and other confining materials have very low transmissivities. Water will tend to flow preferentially where the resistance is lowest; clay is rarely the "preferred" route.

Aquifers existing below confining units are called confined aquifers. They do not recharge vertically to any significant extent as a result of rainfall due to the presence of confining units above them. As a result, these aquifers can have water extracted to a point where there is no additional water available. This is termed "mining." Mining can occur where there is no recharge or where upstream recharge is hindered. Water in such formations completely fills or saturates the formation. Most confined aquifers are under pressure. As a result, they can flow at the surface when the aquifer's static level is higher than that of the top confining layer. This flow is called artesian flow. Such aquifers are referred to as artesian aquifers. Wells drilled into artesian aquifers are called artesian wells. If the water level in an artesian well stands above the land surface, the well is a flowing artesian well. Note that artesian pressure is an indicator that local recharge does not occur.

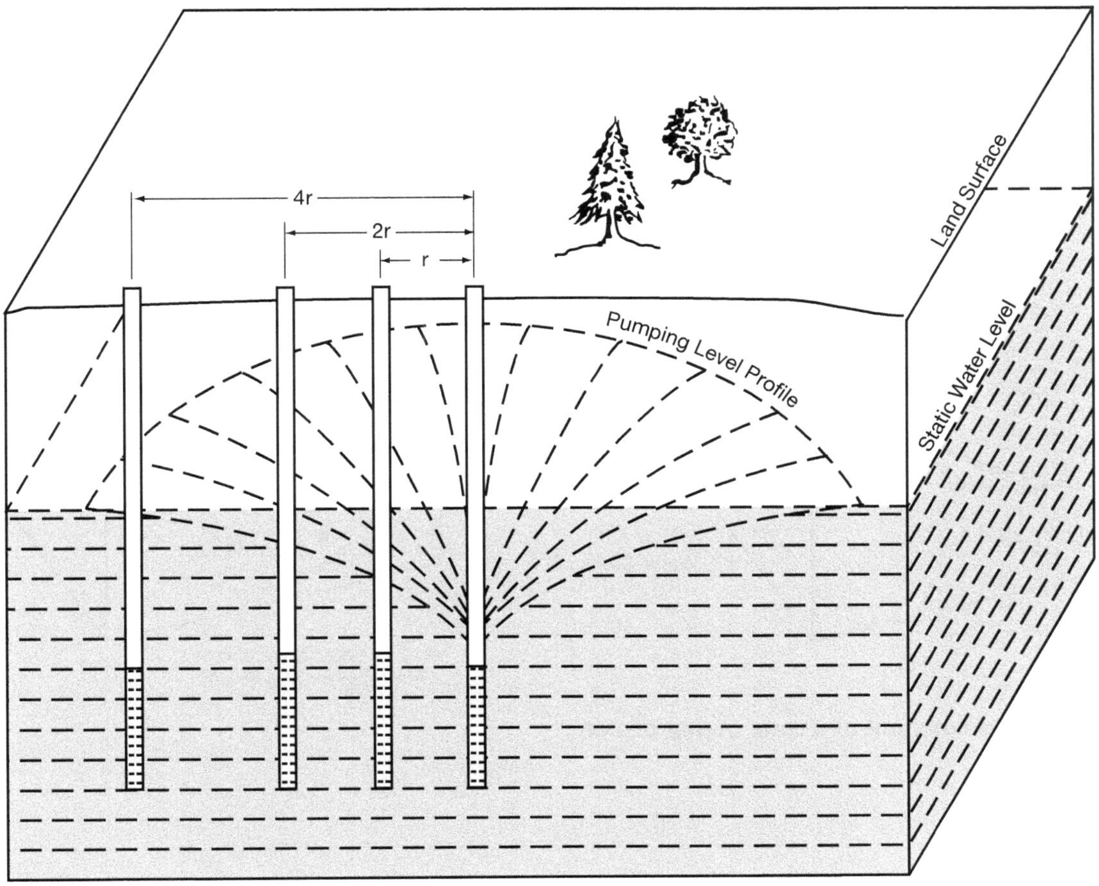

Figure 1-8 Groundwater movement as it relates to topography

In confined aquifers, the pressure is the potentiometric surface. The head relates to the recharge area of the confined aquifer. For example, the Floridan aquifer in South Florida is 1,000 ft (30 m) below the surface. The head is 30 ft (9 m) above the surface. It does not recharge locally. Mapping the aquifer, one notes that the formation tilts to the south. More study reveals that the recharge area extends into southern Georgia. The elevation is between 50 and 100 ft (15 and 30 m). Hence, deducting the head loss in a 400-mi (645-km) distance, the artesian pressure reflects the aquifer recharge zone.

In both confined and water table aquifers, when a well is pumped, the water level falls as the water immediately surrounding the well is drawn through the pumping well. The falling level is called drawdown, and the three-dimensional cross-section is called the cone of depression. If the aquifer transmits water easily (i.e., has high transmissivity), the drawdown is slight, and the cone of depression is flat and widespread as depicted in Figure 1-9, since flowing water can rapidly replenish the pumped water. If the aquifer has low transmissivity, drawdown is significant, and the cone of depression is steep. The higher the rate of flow, the steeper the drawdown and the larger the cone of depression. The cone of depression is also affected by the level of water within the aquifer. Therefore, the type of rock strata that underlies the area has a significant bearing on the quality and quantity of groundwater available and the ability of the mineral material to absorb and store water.

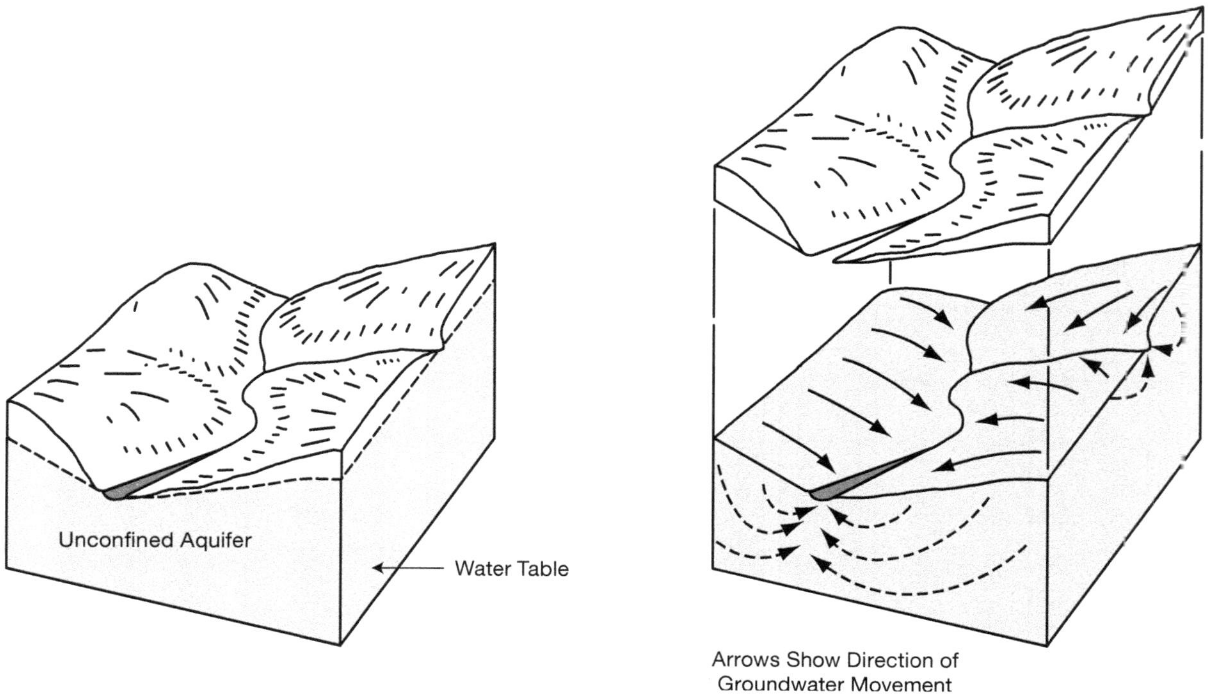

Figure 1-9 Development of a cone of depression

In the western part of the contiguous United States and especially in the alluvial basins region,* conditions are more variable than those described previously. In this area, streams flowing from mountain ranges onto alluvial plains lose water to the alluvial deposits. Thus, groundwater in the upper part of the saturated zone flows down the valleys and at an angle away from the streams.

Pumping Aquifers

As previously noted, more than 41,000 community water systems in the United States use groundwater for their water supplies. The vast majority are relatively small, and their use is limited. A benefit is that water quality and quantity are usually more consistent than surface water supplies. Utilities pumping surficial aquifers will observe the aquifer level falling during dry periods and increasing after precipitation events. That is one benefit of surficial aquifer systems. Pumping protocols can be designed so that the long-term withdrawal rates do not exceed the recharge rates.

The challenge is far more difficult with confined aquifers. Many hydrogeologists assume some degree of recharge occurs through confining units, but studies in many locations indicate that the recharge is overestimated. Lowering pumps exacerbates the problem. The eastern Carolinas suffered this fate in the Middendorf and Black Creek aquifers in the 1990s. The aquifers became depleted, and utilities were forced to seek

* The alluvial basins region occupies a discontinuous area that extends from the Puget Sound–Willamette Valley area of Washington and Oregon to West Texas. This region consists of alternating basins or valleys and mountain ranges.

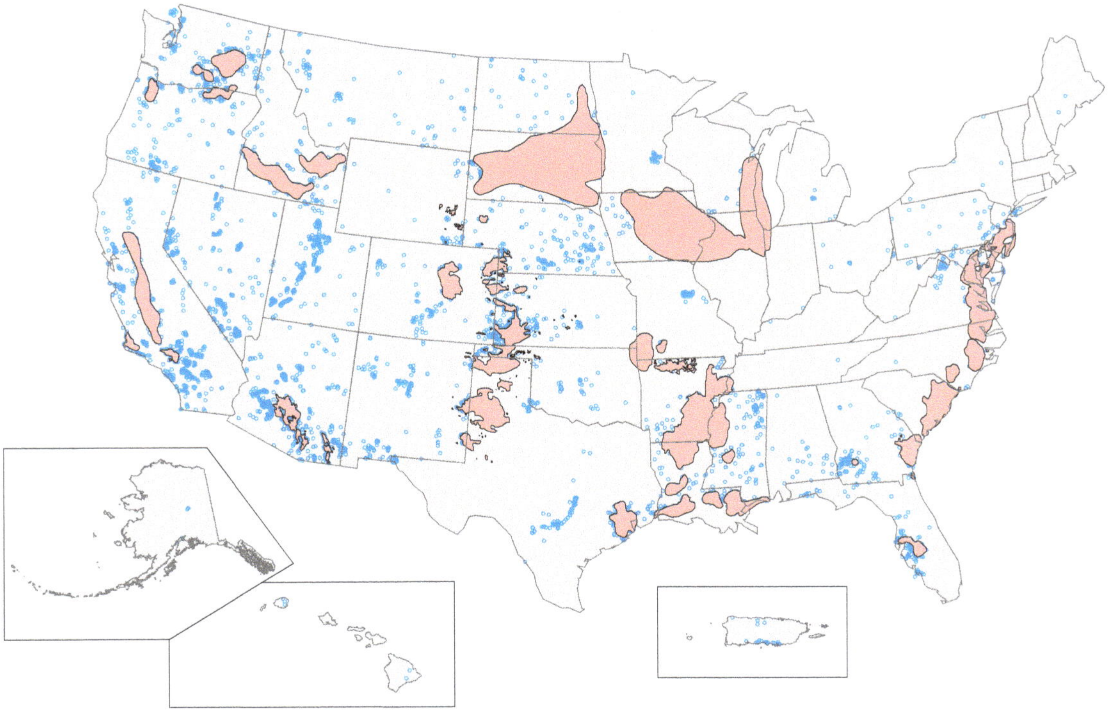

Source: Reilly et al. 2008

Figure 1-10 Water-level declines. Red regions indicate areas larger than 500 mi^2 (1,295 km^2) that have water-level decline greater than 40 ft (12 m) in at least one confined aquifer since predevelopment or greater than 25 ft (8 m) of decline in unconfined aquifers since predevelopment. Blue dots are wells in the US Geological Survey (USGS) National Water Information System database where the measured water-level difference over time is equal to or greater than 40 ft.

other sources—which fortunately were available. However, the dry West presents a larger challenge. Groundwater often recharges at a much slower rate than demands—the Ogallala, Edwards, and Dakota aquifers are examples where agricultural and urban uses have diminished water supplies. In the case of the Ogallala, there is no lower aquifer. Reilly et al. (2008) noted that decreasing aquifers were found across the United States, many in potentially critical areas (see Figure 1-10).

IDENTIFICATION OF SUITABLE GROUNDWATER SUPPLIES

In the past, groundwater was typically assumed not to be affected by surface activities. These perceptions of purity have created expectations of consistent, reliable quality groundwater sources. Unfortunately, groundwater sources cannot be assumed to be safe. In 1969, a national study of 1,000 potable water systems found that more than a quarter of the wells associated with those systems were contaminated by pollutants ranging from minerals to solvents and other hydrocarbons to nuisance microorganisms, sometimes to the point of making groundwater supplies unsuitable for drinking water. These problems led directly to the passage of the Safe Drinking Water Act (SDWA). If suitable, including groundwater supplies and other diverse sources in a water portfolio is an effective way to increase water supply sustainability and reliability. For more extensive discussion on integrated water resources planning, please see AWWA M50. For more discussion on wellfield management and protection from overpumping, please see AWWA M73.

Water Quality Considerations

Traditionally, the materials that have affected the value of a groundwater supply have included naturally occurring minerals, such as dissolved inorganic salts. High quantities of minerals indicate low-quality water. A significant relationship exists among water quality, formation mineral type, and depth of groundwater: the quality of groundwater generally declines with depth as more dissolved minerals occur. Groundwater quality in many sedimentary basins, where the older and deeper sediments were deposited by ancient oceans, can change abruptly with respect to mineral content. Poor-quality water can be drawn upward after production begins ("upconing"). Saltwater upconing can occur even if a production well does not penetrate a saline zone. Similarly, operation of coastal production wells can induce saltwater intrusion into freshwater aquifers through both upconing and lateral movement. Therefore, drilling deeper rarely is a solution to the problem of water quality or decreasing water supply. Knowledge of local hydrology and geology is critical for siting and determining proper well depth and casing.

Today, many forms of contamination exist. Synthetic and naturally occurring organic compounds, gasoline, dry cleaning fluids, pesticides, and heavy metals must be considered when evaluating the development potential of a groundwater resource. Septic tanks are a major pollution source adding microorganisms, minerals, and nitrate to water supplies. Microbial organisms are increasingly a concern. In many cases, construction, maintenance, and operation of facilities to remove these substances is more costly than finding a new water source.

Land Use

Agricultural use of land can affect groundwater quality because of pesticide use, herbicide use, and waste runoff from animal operations. Residential land uses with septic tanks may pollute groundwater with household chemicals, microbiological contaminants, salts, and nitrates. Golf courses are commonly contaminated with arsenic used to control nematodes. Historical land-use practices must be reviewed when siting a well. Potential development and effects of that development should also be considered. The existence of private property rights laws in many states may frustrate utility efforts to protect water sources without significant cost for land acquisition. Wide-scale testing for contaminants throughout a proposed wellfield is required by federal, state, and local regulations. Pollution source areas, especially in capture zones and those that are currently upgradient of the proposed water supply site, are of utmost concern. Because of the types of waste generated within local areas, methods of handling and disposing of the waste, the likelihood of accidental or unreported spills and leaks, and the formation materials, all groundwater supplies should be analyzed for potential pollution contamination. This required testing is regulated for public water supplies under the SDWA but should be done to protect the customer anyway.

As land uses change, an aquifer with near-pristine quality at the time of development may deteriorate in the future. For example, a wellfield might be built in an undeveloped region, but 20 years later, it could be in the middle of a golf course or residential development. Consequently, water resource professionals must obtain and maintain a good understanding of current and proposed urban and industrial growth and land-use zoning of the areas associated with the groundwater supplies. The utility might even need to oppose certain uses that put water supplies at risk—something potentially unpopular with elected officials. Examples are industrial development that displaces irrigated agriculture or paving from urbanization that reduces recharge. Current and ongoing updating of wellhead protection plans can help a utility be aware of land-use changes that may affect water quality.

Today, computer simulation models can depict the long-term effects of a proposed water-well development and provide documentation of any land-use changes that may affect the local hydrology. Computer simulations can be used to account for the lack of certainty about flow patterns and possible changes in flow patterns. In addition, models may indicate the need to alter certain assumptions, such as the flow of discharged material in the unsaturated zone not being governed by groundwater gradients and contaminants moving contrary to prevailing groundwater flow. Such an occurrence created the Walkerton, Ontario contamination incident in 2001. As a result, utilities should be cognizant of groundwater development upgradient or downgradient at an appropriate distance from the well. Testing for pollutants at greater frequency may be appropriate, and establishing early-warning monitoring wells at various depths may be required to maintain groundwater quality. For more extensive discussion on computer simulation models, please see AWWA M73. For more on monitoring wells, see Bloetscher et al. (2022).

Published Reports

Utility-commissioned reports from existing public suppliers, reports from local drillers, and published reports from government agencies such as the USGS are good sources to use when analyzing groundwater conditions. While drillers' logs and well-completion details can be confidential, a utility may be able to negotiate access to these records for specific projects.

Published reports on groundwater resources are available, covering various geographic regions, including major basin areas and state, county, and local regions. For studies within the United States, a good place to begin the literature search is the "Summary Appraisals of the Nation's Ground-Water Resources" (USGS 1978–1982). An individual report covering the region of interest can be obtained separately, or a compilation of all 20 reports is offered by Todd (1983). These reports provide a summary of the quantity and quality of available groundwater. Unfortunately, many of these have not been updated, although USGS constantly writes or updates reports about groundwater availability and aquifer conditions across the United States. By examining the supporting references that are cited for specific localities, data can be arranged into a composite array.

Valuable information regarding contamination sources may be obtained from state and federal programs that administer hazardous waste or effluent discharge permit programs. The agencies administering these programs issue permits to entities that have hazardous chemicals onsite or produce them. They may require information on monitoring of groundwater before issuing permits. Entities that grant operating permits for groundwater use may refer the utility to the records of these entities. Even if they do not, the utility should seek out the records of permits issued and monitoring of same. A local hydrogeologist will likely have a significant amount of information on water quality parameters and drilling conditions.

Caution should be used in any investigation because even site-specific reports are necessarily general in nature, and many local details may be omitted. Local conditions may differ from the regional average, and if good prospects for resource development are rare, one may have to explore areas that do not at first appear attractive in the published reports. Also, water quality can change with time (especially in the shallower groundwater zones), and site-specific variations will not appear in the published literature. Likewise, contamination is unlikely to be noted unless the site has known contamination (e.g., Superfund, state contaminated site inventory) at which time another site should be sought. Therefore, some exploratory site work will have to be completed to provide needed details regarding water quality.

Existing Water Rights

The development of new wellfields should be shown to not infringe on existing water rights or competing water uses. In the United States, both surface and groundwater rights are administered under either the riparian doctrine (water bordering land belongs to land's owner) or the prior appropriation doctrine (first in time, first in right regardless of location). As such, water developers will need to consult local water law and authorities concerning potential infringement on existing water rights.

The water developer should be able to demonstrate "sustainable yield" and also show that the side effects of pumping (such as land subsidence or saltwater intrusion) or interference with competing water or competing land users would be negligible. Sustainable yield means that the amount of water withdrawn is less than the annual recharge, so the water levels will not decline with time. This may be a challenge where confined aquifers do not recharge locally, and the actual recharge areas cannot be studied or controlled by the utility. Confined aquifers are commonly unsustainable, especially in geologically complex areas like the western United States, a concept made abundantly clear by Reilly et al. (2008). Determining whether a water supply is sustainable may be accomplished via computer modeling by using a program such as USGS's MODFLOW program. Local and state USGS branch offices might have staff that can assist with modeling. These models also can be used to demonstrate to regulatory agencies that proposed drawdowns will not be detrimental to the environment or adjacent users or exceed the drawdowns set by regulation. Computer programs of varying complexity, as discussed in M73, are also available for use in conjunction with water quality analyses.

Changes Affecting Evaluation

Changes in local economies may influence local officials to develop more groundwater resources or the public to demand higher-quality water. Finding more water may require use of poorer water sources, and drilling new wells may challenge the utility provider's aquifer protection programs or make public relations efforts difficult when trying to restrict land uses. Consequently, reliable quality measurements must be obtained when evaluating any groundwater system.

Deterioration of water quantity and quality can have grave consequences for those dependent on the water resource in the form of higher treatment costs or abandonment of the wellfield. However, changes other than those measured in the field can often be of great concern. Perceived changes may be caused by the availability of improved analytical methodologies (which detect levels of compounds not previously measurable or of emerging contaminants); regulatory priority shifts toward or away from protection of natural resources, including aquifer classifications; new toxicological data or reinterpretation of existing toxicology; or the integration of facts regarding groundwater quality related to the safe use of all water resources. Findings of perfluoroalkyl and polyfluoroalkyl substances (PFAS), perfluorooctanesulfonic acid (PFOS), and pharmaceuticals are examples of contaminants that have been of concern lately, largely because detection methods are now available to find tiny quantities of these constituents. Utilities need to monitor for regulated compounds of the PFOS family as a result of new 2024 regulations.

LAND SUBSIDENCE

A significant consequence of groundwater development can be downward movement of the land surface called subsidence. Since water is an uncompressible fluid, it maintains the formation in some situations. When the water is removed from the formation, the pressure is relieved, and the underlying rock can collapse. When the formation collapses, it causes the surface topography to be altered. Development of groundwater needs to include consideration of possible land-surface subsidence. Sinkholes are a localized example of this phenomenon.

Estimating Subsidence

Subsidence is often caused by the extraction of water from aquifers. The most readily available of the necessary data is the amount of compressible material in the formation. Such data may be obtained from evaluation of drilling logs of test wells. Data on water-level changes can be used to make estimates of pressure change (stress change) at various depths for various time intervals. Where subsidence has been well documented, subsidence data may be coupled with the amount of compressible material to determine compressibility. Unfortunately, the information needed is not often available in sufficient detail in most areas.

The magnitude of subsidence should be estimated, especially in aquifers with high clay or sand content or where prior subsidence has been noted. Subsidence along faults (activation or acceleration) that could lead to structural damage must be estimated. Data on the degree of compressibility of the subsurface material are less readily available than either of the other factors used to predict subsidence. Laboratory values of compressibility determined from tests of cores have been used with limited success because of the expense of obtaining undisturbed or authentic cores and the difficulty in obtaining representative cores precludes their use for regional appraisal.

GROUNDWATER TERMS

The following terms are used to explain the concepts found in this chapter and throughout the remainder of this manual:

Anisotropic aquifer is one in which the aquifer formation does not transmit water equally in all directions (i.e., the horizontal and vertical permeability are not equal, causing the water to move preferentially in one direction with respect to the other two).

Artesian aquifers occur when the aquifer's static level is higher than that of the top confining layer.

Aquiclude is a low-permeability geologic unit that forms either the upper or lower boundary of a groundwater flow system. Clay is a typical aquiclude material. This term is rarely used in the field today but may exist in older hydrogeologic reference materials.

Aquifer is a geologic formation, or part of a group of formations, which is saturated and sufficiently permeable to transmit economic quantities of water to wells and springs.

Aquifer storativity is the volume of water added to a unit horizontal area of the aquifer per unit rise in the water table elevation (i.e., the ability of the aquifer to increase its capacity to store water).

Aquitard is a low-permeability geologic unit that can store groundwater and transmit it slowly from one aquifer to another. A leaky aquifer formation may be an aquitard. This term is rarely used in the field today but may exist in older hydrogeologic reference materials.

Effective pore space is an indication of how much of the void space within the rock or soil is capable of transmitting water. This is important because some rock formations may have considerable pore (or void) space, but the rock or soil may have difficulty transmitting water because the pores are not interconnected.

Evaporation is the transformation of water from the liquid phase to a vapor phase in the atmosphere.

Evapotranspiration is the combination of evaporation and transpiration. Water less than 4 ft below the ground surface may be subject to evapotranspiration.

Groundwater is water contained in interconnected pores located either below the water table in an unconfined aquifer or in a confined aquifer.

Hydraulic conductivity is a measure of the ease with which fluid is transported through a porous matrix (also called coefficient of permeability).

Hydraulic gradient is a measure of the vertical change of the aquifer over a given distance from a stationary base (i.e., the slope of the aquifer formation).

Hydrologic cycle is the movement of water in the environment wherein water falls to the surface as rainfall, runs off over land to water bodies and/or infiltrates into the groundwater via percolation, returns to the atmosphere via evaporation/transpiration, and then returns as rainfall. This is best demonstrated in Figure 1-1.

An ***isotropic aquifer*** is one in which the formation transmits water equally in all directions within the aquifer formation.

Piezometric surface is the level to which water will rise in an aquifer under natural conditions. In a water table aquifer, the piezometric surface is the water table level. In a confined aquifer, the piezometric section may be significantly above the top of the rock formation; the aquifer is under pressure. Artesian wells are confined aquifers with the piezometric level above the ground surface, meaning that water could flow to the surface unaided.

Potential evapotranspiration is the amount of evapotranspiration that would occur, assuming the soil moisture was adequate at all times. During drought conditions this will not occur, so actual evapotranspiration may be less. Because of the disparity between the rainfall throughout the year, the actual evapotranspiration in the winter months is less.

Specific discharge potential is the amount of water that an aquifer could discharge given its hydraulic conductivity and piezometric head.

Surficial reflects the location of the aquifer—a surficial aquifer has a connection directly to the surface. Therefore, rainfall can recharge a surficial aquifer as opposed to a confined aquifer, which is not recharged by rainfall.

Transmissivity is the aquifer characteristic that is defined by the rate of flow per unit width through the entire thickness of an aquifer per unit gradient. This is valid only in two-dimensional flow.

Transpiration is the process in which plants absorb water from the soil and release it to the atmosphere through their leaves.

Viscosity combines with the fluid density to form kinematic viscosity, which is important in determining the hydraulic conductivity of the soil or rock matrix.

Void space is the pore space within the soil or rock matrix that does not contain solid material.

REFERENCES

American Water Works Association (AWWA). 2015 (1st ed.). M63, *Aquifer Storage and Recovery*. Denver: AWWA.

AWWA. 2021 (1st ed.). M73, *Groundwater Management*. Denver: AWWA.

Bloetscher, F., A. Muniz, and J. Largey. 2022 (2nd ed.). *Water Supply Wells: Siting, Drilling, and Construction*. Denver: AWWA.

NOAA (National Oceanic and Atmospheric Administration). 2021. US Climate Extremes Index. Accessed January 2021.

Reilly, T.E., K.F. Dennehy, W.M. Alley, and W.L. Cunningham. 2008. "Ground-Water Availability in the United States." Reston, Va.: US Geological Survey.

USGS. 1978–1982. "Summary Appraisals of the Nation's Ground-Water Resources." Prof. Paper No. 813, Chap. A–U. Denver: USGS.

Todd, D.K. 1983. *Ground-Water Resources of the United States*. Berkeley, Calif.: Premier Press.

This page intentionally blank.

Chapter 2

Wells: Types, Construction, and Use

Many methods of well construction have been developed, and several different construction types are generally used in most jurisdictions. In many cases, certain standards are created. AWWA has one such standard: ANSI/AWWA A100 Water Wells. Proper well construction should be based on a thorough engineering study and design to best accommodate existing conditions or requirements.

TYPES OF WELLS AND THEIR CONSTRUCTION

In the water industry, "well type" generally refers to the method of well construction because most wells are used for water supply or monitoring purposes. Examples include dug, bored, driven, or drilled. A fifth type of well, which is not named for its method of construction but rather a description of its configuration, is the radial collector well. An emerging concept is the use of horizontal wells without collector caissons for water supply. Each type of well has advantages, such as ease of construction, storage, capacity, ability to penetrate various formations, and ease of safeguarding against contamination.

Dug Wells

This type of well technology is rarely used in the United States and Canada, although it may be appropriate for developing countries.

A dug well can furnish relatively large supplies of water from shallow sources. The yield from a dug well increases with diameter, but the increased yield is not proportional to the increased size. A dug well is of large diameter, usually 8–30 ft (2–9 m), when installed for municipal purposes, with a depth varying from 20–40 ft (6–12 m). Because of the large opening, dug wells are easily polluted by surface water, airborne material, and objects falling into or finding entrance into the well.

Construction. Generally, a dug well is circular because this shape adds strength and is usually easier to dig. Material is frequently excavated using a pick and shovel or a hoist with a bucket. Clamshell buckets with power hoists can be used when no large boulders or thick layers of clay or hardpan are encountered.

If the formation in which the well is being dug will stand without support, lining the excavation may not be necessary until the water table is reached. Below the water table, sheet piling temporarily braces the sides of the excavation. Later, after the lining or casing (usually called the curb) is placed, the piling is removed. Some older wells are brick-lined.

To minimize surface pollution, monolithic concrete curbs are used. The curbs are built in rings 3–4-ft (1–1.2 m) high. As the well depth increases and the curbs sink, additional rings are added. The rings are reinforced and are tied together by vertical steel. Both inside and outside forms are used to get a smooth surface that will sink easily. The portion of a curb that lies in the water below the limit of drawdown should be perforated to permit water to flow horizontally into the well. Short pieces of pipe are cast in the curb, with several in each square foot (0.09 m^2). Graded gravel should be placed around the outside of the curb to keep sand from coming through the perforations. Pipe and gravel sizes depend on the natural formation grain size. Gravel should also be placed in the base of the excavated well. In some cases, precast concrete pipe sections can be used as the well casing; the pipe below the water table must be perforated.

Bored Wells

A bored well is installed where speed and economy are important and where water can be reached at shallow depths through unconsolidated formations. An auger can be used only where formations, although relatively soft, will permit an open hole to be bored to depths ranging from 25–60 ft (8–18 m) without caving. The most suitable formations for bored wells are glacial till and alluvial valley deposits. Generally, this type of well is not often used for municipal water supplies.

Construction. A bored well is constructed using either hand or power augers. The same type of hand auger used for digging shallow holes can be used for well construction. However, extensions are needed for the auger to handle the greater depths encountered in well construction. Power-driven augers are half-cylinder, open-blade, or cylindrical-bucket-types with cutting blades at the bottom. The material cut by the blade is collected in buckets lowered into the hole and then removed. Bucket-type augers can be used to construct wells up to 36 in. in diameter and can be used where large boulders are present. This method of drilling is commonly used for short-term wells, such as for construction dewatering, or for recovery wells for cleaning up groundwater contamination.

Because sand and gravel are encountered below the water table, the well casing is lowered to the bottom of the hole. Boring continues by forcing the casing down as the material is removed from the hole. After the well is completed, the annular space between the borehole and the outside of the casing should be filled with cement grout to prevent the supply from becoming contaminated.

Driven Wells

Driven wells are simple to install and economical but practical only where the water table is shallow. A driven well consists of a pointed screen, called a drive point or well point, and lengths of pipe attached to the top of the drive point. The drive point is a perforated pipe covered with woven wire mesh, a tubular brass jacket, or a pipe with perforations that are similar to screens for drilled wells and adaptable to driving. A pointed steel tip at the base of the drive point breaks through pebbles and thin layers of hard material and opens a passageway for the point. A driven well varies from 1¼–4 in. (32–100 mm) in diameter and is a maximum of about 30–40 ft (9–12 m) deep.

For municipal water supplies, the driven well is used where thin deposits of sand and gravel are found at shallow depths. The production rate of these formations is limited, and a single well does not produce sufficient water for more than a few people. However, a battery of well points, with the wells located a reasonable distance apart and connected by a common header to the pump, could develop sufficient water to supply a small community. In this case, a suction pump raises water in the wells to a point at which it can be pumped into the distribution system. This configuration is typically limited to areas where the water table is within about 15 ft (4.5 m) of the surface. Driven wells are also used as observation wells during aquifer tests.

Construction. When constructing a driven well, an outer casing is first installed. The outer casing protects the inner casing to which the pump is attached. A partial vacuum occurs in the inner casing that can draw contamination at leaking joints. The outer casing is usually 2-in. (50-mm) larger in diameter than the well casing. It should extend a minimum distance of 10 ft (3 m) below the ground surface.

In sand and gravel, the outer casing should extend to just above the drive point. The outer casing can be driven with a sledgehammer. A tripod and pulley can be used, which raises and lowers a heavy block onto a drive cap placed on top of the casing. Extra heavy pipe should be used to withstand the load. The sand and gravel in the outside casing are removed by an auger during driving. If the ground is clay, the outside casing should be set in a hole prepared with an auger. Under such conditions, a 10-ft (3-m) depth usually affords sufficient protection. After the casing is set, the annular space between the borehole and the outside of the casing should be sealed with cement grout.

The next step is to lower the drive point, attached to the bottom of a string of inner casing, into the hole. The drive point is driven below the bottom of the outside casing to the water-bearing formation. This depth may be tested periodically by pouring water down the well casing and observing how quickly the water moves down the well. This can be accomplished by attaching a pump or by jetting water down inside the casing to wash out the screen. When the final depth has been reached, raising or lowering the pipe 1 ft (0.3 m) or so frequently brings a greater portion of the screen into contact with the water-bearing formation, resulting in increased production.

Wellpoints to dewater construction sites are common uses for driven wells.

Drilled Wells

A drilled well is commonly used for municipal supply and can develop water from shallow or deep sources in unconsolidated sands and gravels or rock. Pipe diameters range from 2–48 in. (50–1,210 mm) and are larger for drilled gravel-wall wells. High-capacity wells that can produce thousands of gallons per minute can be developed, provided that the aquifer can support these production rates.

Construction. Well drilling is a complex and specialized construction task and is only summarized here. More complete discussions are provided in standard industry references such as Bloetscher et al. (2022), NGWA (1998), ADITC (1997), Roscoe Moss Co. (1990), and Driscoll (1986). When constructing a well, a drilling rig is used to excavate or drill a hole, and then a casing is forced or placed in the hole to prevent it from collapsing. When a water-bearing formation of sufficient capacity is reached, a screen is set in place that allows water to flow into the casing and holds back the fine material in the formation. When the drilled well passes through rock, a screen is usually not used unless the formation is fractured.

Wells are drilled using many different types of equipment. The methods used to construct wells are classified as cable-tool, rotary, reverse-circulation rotary, California, and jetting. Each method is briefly described in the following text.

Cable-tool method. The cable-tool method of drilling is used extensively for wells of all sizes and depths. Cable-tool methods are also called percussion, spudder, and solid tool. The details for constructing and operating the drilling machines vary widely, although all machines dig a hole using the percussion and cutting action of a drill bit. The bit is at the end of a string of solid drilling tools. The drilling tools are placed at the end of a cable that is alternately raised and dropped. The drill bit, a club-like, chisel-edged tool, breaks the formation into small fragments, and the reciprocating motion of the drilling tools mixes the loosened material into a sludge or slurry. Note that the action of the bit can damage the borehole wall, resulting in a lower well efficiency than desired.

Generally, several feet (1–2 m) of hole are drilled at each run of the drill tools. After each run, the tools are pulled from the hole and swung aside while a bailer is used to remove the slurry. The bailer consists of a 10–25-ft (3–8-m)-long section of tubing with a check valve in the bottom. The bailer is smaller in diameter than the drill hole so that it can move up and down freely.

An experienced driller adjusts the length of the drill cable so that the bit will strike with the right amount of weight and stroke. The driller holds the cable and feels the jarring when the cable is dropped, which indicates how well the tools are operating. The driller adjusts the length of stroke and rapidity of blows according to the cable vibrations.

Drill hole characteristics. The drill hole must be straight and vertical or plumb. Usually, the first indication that the hole is out of plumb is that the drilling tools begin to stick. If this occurs, drilling should stop, and the hole should be realigned. A straight, vertical hole permits the lowering of a pump to the desired depth and prevents damage to pumping equipment. Although pump manufacturers state that pumps will operate satisfactorily when slightly off vertical, a well out of alignment causes severe wear on the pump shaft, bearings, and pump column pipe. In extreme cases of misalignment, lowering a pump into a well or pulling it out may become impossible.

The drilling contractor should be required to verify, through testing, that the completed hole is straight and plumb. In one such test, a 40-ft (12-m) length of pipe, often called a "dummy," is lowered into the well to the depth of the lowest anticipated pump setting. The outer diameter of the test cylinder should not be more than ½-in. (13-mm) smaller than the diameter of the casing being tested. The test cylinder should move freely throughout the length of the casing. The well should not vary from the vertical by more than two-thirds of the inside diameter of the well per 100 ft (30 m). See ANSI/AWWA A100-06 for details.

After the hole has been drilled to full depth and verified, a well screen section is lowered inside the drill casing until the screen is even with the aquifer. The drill casing is then retracted a distance almost equal to the length of the well screen to expose the formation materials to the well screen.

Formation recognition. To determine the exact kind of material being drilled, samples should be taken of the cuttings at 5-ft (1.5-m) intervals or at each noticeable change in formation. Generally, water-bearing formations are easily detected. A sudden rise or fall of the water level in the well frequently indicates that a permeable formation has been encountered. Often, water is added inside the drilling pipe to maintain sufficient head to prevent excess formation materials from heaving and washing inside of the casing during drilling. Sand, gravel, sandstone, and limestone formations produce the largest quantities of water. Drill operators need to be especially watchful when drilling in these formations.

Advantages. With the cable-tool drilling method, a more accurate sample of the formation is obtained because only rock and the aquifer water are in the borehole. As a result, the quantity and quality of each formation can be determined and less water is necessary for drilling operations because drilling mud is not used during drilling, which minimizes plugging of the formation and simplifies development of the well. In most cases, a cable-tool rig is light and can easily traverse rough country.

Rotary method. In the rotary method of well drilling, the hole is made by rotating a bit located at the end of a string of drill pipe (see Figure 2-1). This method is sometimes augmented by a percussion action, with cuttings removed by a circulating fluid. The rotation speed of the drill pipe and the bit and the type of fluid used (and its flow direction) can vary by type or form and size of bit, the characteristics of the formation to be drilled, the strength and weight of the drill pipe, and the size and depth of hole to be drilled. The bit and drilling speed should be selected so that cuttings are not produced faster than circulating fluid (mud) can carry them to the surface. If the hole diameter becomes greatly enlarged, the uphole velocity may be insufficient to remove the cuttings from the well.

The drill pipe is hollow (see Figure 2-2); therefore, fluid can be pumped to the bit. When drilling in unconsolidated or cemented granular formations (clay soil, sand, gravel, weak sandstone, and shale), a "mud" or fluid of sufficient viscosity is used to lift cuttings to the surface. The fluid must also have the necessary sealing qualities and weight, which help stabilize the well bore. The fluid may be formed by normal drilling operations, starting with clear water; however, an engineered mud fluid is preferred for the best results.

The fluid is usually prepared in a large tank or pit located near the drilling rig. The prepared fluid circulates through the drill pipe and out through holes in the bit, where it sweeps under the bit, picks up the material loosened by it, and carries it up the borehole to the surface. In mud rotary drilling, the fluid from the well flows into a settling tank or pit, where the cuttings settle out. The fluid, now free from coarse materials, flows into another pit, where it is picked up by the pump for recirculation. Formation samples are typically collected by sampling the drilling mud as it exits the borehole before it reaches the settling pit (see Figure 2-3). Clays at the site can be used, but commercial colloidal material (bentonite clay base), which is purchased in powdered form and mixed with water, is preferred. Specialized polymer muds are also available for certain drilling conditions where bentonite mud is not preferred.

For rotary drilling, the mud can go down through the drill pipe or up through the drill pipe. Air-based fluids are also used, principally in rock drilling. The fluid may be air only, pressurized by a compressor, or air mixed with water (mist) or polymers (foam). Foam permits vastly improved cuttings clearance and is used where the fluid system (compressor and drill tools) capacity would not be able to generate enough uphole velocity to clear cuttings.

Down-the-hole (downhole hammer, hammer drilling). The down-the-hole method involves a pneumatically operated bottom-hole drill that efficiently combines the percussion action of cable-tool drilling with the turning action of rotary drilling (NGWA 1998, ADITC 1997). The pneumatic drill can be used on a standard rotary rig with an air compressor of sufficient capacity. It is used for fast and economical drilling of medium to extremely hard formations. Fast penetration results from the blows transmitted directly to the bit by the air piston. As in air rotary, air circulation flushes the bit and carries cuttings to the surface. Air also powers the hammer. The system both hammers and rotates the tungsten-carbide bits against the borehole face to dislodge cuttings.

Continuous hole cleaning exposes new formation to the bit, and practically no energy is wasted in redrilling old cuttings. As the force to drill the formation is applied by the piston at the bit surface, the down-pressure necessary for air rotary hole advancement is not needed, and holes are typically straighter.

Down-the-hole drilling is generally the fastest method of penetration in hard rock. The bit is turned slowly (5–15 rpm) by the same method by which the drill bit in the fluid or air drilling operation is rotated. Foaming additives are occasionally used to increase the uphole carrying capacity of the return air.

Source: Courtesy of Frederick Bloetscher

Figure 2-1 Example of drill rig with rotary bit

Source: Courtesy of Frederick Bloetscher

Figure 2-2 Hollow stem drill pipe

Source: Courtesy of Frederick Bloetscher

Figure 2-3 Example of a mud pit

Reverse-circulation rotary method. The reverse-circulation rotary method differs from the straight rotary method in that the fluid circulates in the opposite direction. The pit is constructed so that the drilling fluid will flow down the borehole and rise in the drill pipe, carrying the cuttings with it. A high-capacity (500 gpm [1,900 L/min] or greater) pump is attached to the drill pipe and keeps the fluid moving at high velocity. The pump may discharge to waste if a large, fresh supply of water is available, or the cuttings may settle, and the fluid recirculates.

The use of organic polymers reduces the detrimental effect of mud cake in the bore. The reverse-circulation method primarily uses clear water with no mud additives. Keeping the borehole open requires a large volume of water to maintain a head above the natural static water level. The head results in a flow into the formation from the bore, rather than the reverse, and prevents the wall from caving. When substantial thicknesses of materials that will not accept water are encountered, caving may result from the wash action of the fluid moving down the hole. The reverse-circulation method is particularly well-suited for artificial-gravel-pack wells because less mud-cake forms on the face of the bore and into the formation, reducing development time. If the formation is highly permeable and the water supply at the surface is limited, processed clays may be added ("mud the hole"), which creates a plastered hole similar to the straight rotary process. This step defeats the primary advantage of the reverse-circulation method of construction. In deeper wells being drilled with the reverse-circulation method, an air line can be added inside the drill pipe to aid in maintaining flow circulation. This also compensates for the increased viscosity of the fluid containing the formation cuttings and allows drilling to depths greater than 1,000 ft (305 m).

Typically, reverse-circulation methods can drill larger holes than direct rotary methods—up to 5 ft (1.5 m) in diameter. A test well can be drilled and, if desired, the hole can be abandoned at a minimum expense by plugging. Proper plugging cannot be performed with the casing in the hole unless the casing is perforated to get the sealant between the casing and the borehole wall. In soft, loose, unconsolidated materials such as dune sand and quicksand, the hole may be difficult to keep open unless the rotary method is used. It also is generally faster than cable-tool drilling for drilling wells greater than about 18 in. (45 cm) in diameter.

Dual-rotary. Dual-rotary drilling uses a drilling rig with two rotary drives. One drive is typically used as a casing driver to rotate the outer drilling casing into the ground. The casing can be fitted with a hardened drive shoe or cutting shoe to help the casing penetrate rock or unconsolidated formation materials. The second drive is a standard rotary drive mounted on the drilling rig mast that uses air or fluid rotary drill pipe and bits.

The drill pipe is hollow, which allows air or water to be pumped down to the drill bit. The drill cuttings are carried up between the drill pipe and the outer driven casing to the surface. The cuttings and drill fluid are separated, and samples of the formation can be collected. The drilling fluid can be recirculated back to the drill bit. Drilling mud additives are typically not required, which eliminates the plugging of the formation and borehole, and simplifies well development.

Dual-rotary drilling equipment is common for drilling wells for both municipal water supply and in environmental applications. The main reasons are drilling mud is not needed and zones of undesirable water quality can be isolated or "cased off."

California method. The California, or stovepipe, method of well construction was developed in California mainly for sinking water wells in unconsolidated alluvial materials of alternate strata of clay, sand, and gravel. Wells 16–20 in. (400–500 mm) in diameter and up to 300 ft (90 m) in depth are constructed using this method. The California drilling method uses the same general principles used in the standard cable-tool method except that a specially designed bucket is used as both bit and bailer. Short lengths of sheet metal, either riveted or welded together, are used for casing.

The mud-scow bit used in this method consists of a disk valve bailer with a sharp-edged cutting shoe on the bottom. Similar to an ordinary sand bucket, it is heavier, larger in diameter, and has the cutting shoe on the bottom. Each time the bit is dropped, some parts of the cuttings are trapped in the bailer. When filled, the bailer is pulled to the surface and emptied.

At the bottom end of a string of California stovepipe casing is a riveted steel starter 10–25-ft (3–8-m) long and made of three thicknesses of sheet steel with a forged steel shoe at the lower end. This reinforcement prevents the bottom from collapsing when under pressure. Above the starter, the casing consists of two sizes of sheet steel made into riveted or welded lengths from 2–6 ft (0.6–2 m). The larger size casing fits snugly over the smaller size. Each outside section overlaps the inside section by half its length so that a smooth surface results both outside and inside when the casing is in place. In this way, the inner and outer joints never coincide.

The casing is forced down, length-by-length, by hydraulic jacks anchored to two timbers buried in the ground. These jacks press on a suitable head attached to the upper section of the stovepipe casing so that the end of the casing will not be telescoped. The casing may also be driven by raising and lowering the tools with a driving head.

After the casing is in place, it is perforated using a Mills knife or similar device that tears the metal. The openings must not be too large, and the pipe must not have too much area perforated.

Caisson well construction. This type of well construction is used in shallow, very loose, and permeable alluvial formations where fluid losses may exceed the capacity of the drilling water supply and where no drilling mud is to be used in the drilling process. The hole is made by bailing and sinking a casing very large in diameter to about 15 ft (4.6 m). The next casing is installed in a concentric manner, one size smaller in diameter (6 in.) than the first, using the same bailing method until it is extended another 15 ft (4.6 m) below the first casing. This process is continued until the bottom of the desired formation, or the desired depth is reached. The last casing installed should be of the minimum diameter of the borehole designed or specified.

The well screen and casing are installed concentrically in the temporary inner casing, and placement of gravel in the annular space is initiated. As the gravel material is placed, the temporary inner casing is withdrawn, keeping the gravel pack material about 6.5 ft (2 m) above the bottom of the casing during withdrawal. This process continues until the gravel pack extends several feet above the well screen. A grout seal should be installed around the permanent well casing, and the remaining temporary casings should be removed and seals installed in a similar manner. Because no fluid movement or drilling additives are used in the construction, the resulting well efficiency is high and little well development is generally needed.

Jetting method. Jetting is used to drill a vertical well when water is found in sand at shallow depths. It can also be used for deep wells. Jetting equipment consists of a drill pipe or jetting pipe equipped with a cutting bit on the bottom end. Water is pumped into the well through the drill pipe and out of the drill bit against the bottom of the drill hole.

The casing is usually sunk as drilling proceeds. In some instances, the casing will sink a considerable distance under its own weight. Ordinarily, however, a tripod and drive weight are needed to force it into place. As a rule, one size of casing is used for the entire depth of the well. However, if a well is deep, driving a single string of casing to full depth can be difficult. Often, several strings of casing of different diameters are telescoped one inside the other to reach full depth.

After the casing is lowered to the water-bearing formation, the well screen and pipe are lowered into the casing. The outside casing provides protection to the inner casing connected to the screen. The well screen is exposed to the water-bearing formation by pulling back the outer casing a distance equal to the length of the screen (similar to the cable-tool method of drilling).

Certain conditions can make this method of well construction difficult. Rock formations and boulders are barriers that cannot be overcome. Formations of clay and hardpan are other types of materials that can present problems.

Other construction methods. Other types of construction include under-reamed and bail-down methods. The under-reamed, gravel-wall well is drilled to the top of the water-bearing formation, and the casing is set. The formation is then under-reamed, and the gravel and screen section are placed. The bail-down method makes use of a cone-bottomed screen. The outer casing is sunk to the top of the water-bearing formation, and the screen is bailed into place while gravel is fed into the space between the inner and outer casing. In the pilot hole method, the gravel is fed into the formation through small pilot wells evenly spaced around a central well. The fines are withdrawn through the central well.

Another type of well construction common to the mid-Atlantic and Gulf Coast states is the two-piece well construction. A large-diameter hole is drilled by direct rotary methods to the top of the formation to be penetrated by the well screen. The outer casing is grouted to the borehole wall. The formation to be developed is drilled with an under-reamer bit to a diameter larger than the casing installed. The well screen and inner casing (or lap pipe) are installed, and its gravel pack material is installed by a tremie pipe. In deep wells, the inner casing must lap the outer casing with sufficient length to overcome the buoyant lift forces when the well pump is above the lap pipe. Typically, this length is a minimum of 50 ft (15.2 m) with a maximum of 90 ft (27.4 m). A short section of well screen can also be installed 5 ft (1.5 m) above the regular screen to relieve the buoyant pressure in the gravel pack between the casings.

Radial Wells

The radial, or horizontal collector, well is widely used because it can produce very large quantities of water. In many cases, a radial well is located along the shore of a lake or river because infiltration from the water body can recharge the well.

A radial well is essentially a combination dug well and a series of horizontally driven wells projecting outward from the bottom of its vertical walls. The main well, or central caisson, serves as a collector for the water produced from the individual horizontal wells called laterals. The laterals are installed in coarse formations, often in more than one layer or tier. The lateral well screens can be installed, and a natural gravel pack developed, or an artificial gravel-pack filter can be constructed around the well screens to accommodate finer-grained formation materials. A general schematic diagram of a radial collector well is shown in Figure 2-4.

Construction. The central caisson of a radial well is constructed of reinforced concrete. It has an outside diameter of 12 to more than 20 ft (4–6 m) and an inside diameter ranging between 9 and 20 ft (3–6 m). The wall is generally 12–24-in. (300–600-mm) thick and is poured in circular sections 8–12-ft (2–4-m) high. The bottom of the first section, or ring, is formed with a cutting edge to facilitate the caisson's settling in the excavation and to provide a stronger bearing surface for the base of the caisson shaft. Wall ports are usually cast into the first section of the caisson, which are then used to direct the installation of the horizontal well screens.

Material is excavated from within the caisson, keeping the caisson as plumb as possible. Each section is keyed and tied to the previously poured unit for structural stability and watertightness. The final depth of the cutting edge is usually at or several feet (1 m) below the bottom of the water-bearing formation. When the caisson has been sunk to the design elevation, a concrete plug is poured in the bottom.

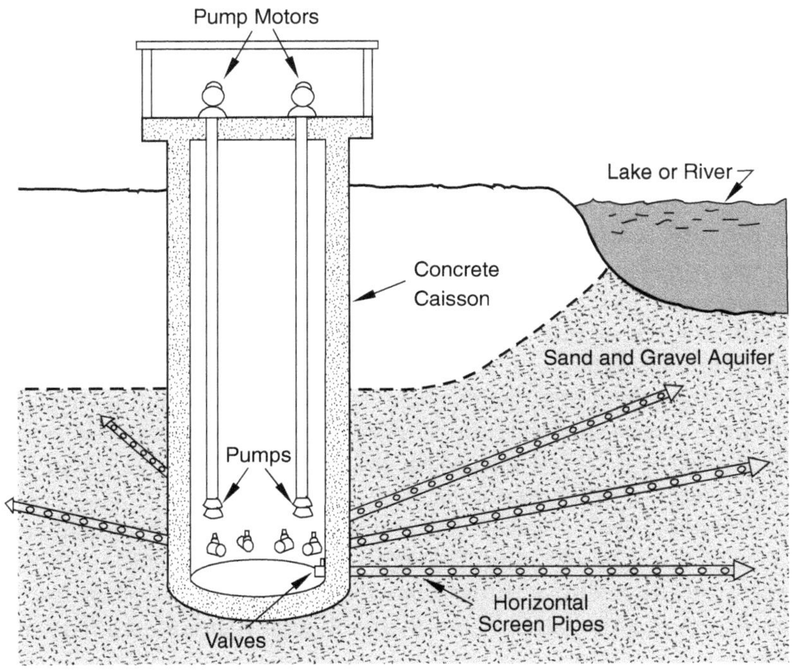

Figure 2-4 Details of a radial well

Laterals are projected horizontally through wall sleeves from the central caisson into the water-bearing formation. The laterals are constructed of slotted or perforated pipe or conventional wire-wound well screens. They are generally positioned near the bottom of the water-bearing formation. In some formations, they may be placed on a selected horizon with efficient hydraulic characteristics. The entire length of the lateral is perforated with the exception of a 5.10-ft (1.53-m) blank section extending from the outer wall of the caisson. A gate valve is installed on each lateral inside the caisson to make it possible to cut off the flow into the caisson for dewatering. The caisson is typically extended above known or anticipated flood elevations. A superstructure is erected on top of the caisson for housing the pumps, piping, and electrical controls.

Advantages. The radial well offers several advantages over conventional vertical production wells. A much greater length of well screen can be installed at a given well site because the well screen is typically placed near the base of the aquifer and is not limited to the saturated thickness of the formation. This greater length of screen results in extremely low entrance velocities through the well screen openings, reducing the rate of well screen plugging and typically reducing well maintenance. The location of the laterals near the base of the formation also allows a greater saturated thickness of the aquifer to be used so that higher yields can be obtained at an individual well site. A higher yield allows a single radial well to replace multiple conventional vertical wells, reducing the number of pumps and well systems to be maintained.

Gravel-Packed Wells

A "gravel-packed well," also called a gravel-wall well, is not really a type of well but involves construction where a gravel pack is installed between the formation and screen to keep fines and sand out of the well. The concept uses larger slot sizes in the well screen section than would be possible if the area surrounding the screen were not gravel packed. The amount of open area in the screen is increased, as is the effective diameter of the well. Also, the amount of fine sand from the water-bearing formation entering the well is reduced. Lower entrance velocities at the screen openings and increased flow per unit of head loss are also achieved.

A gravel-wall well must be carefully designed. The material used in the gravel filter must be clean, washed gravel composed of well-rounded particles. The filter size depends on the size of the natural formation and the intended slot openings of the well screen. Without proper gravel size, the fine sand will not be kept out, and the yield will be adversely affected. The size of individual grains of gravel filter material should be four to six times larger than the median size of the natural material. At the same time, the uniformity coefficient* of the gravel treatment should be similar to that of the formation material. The slot size for the screen should retain 90 percent of the pack material. An artificial gravel-pack filter can also be installed around the lateral well screens in a radial collector well to match finer-grained formation materials.

Construction. Selected gravel is placed between the outside of the well screen and the face of the water-bearing formation (the drilled borehole). This method is especially useful when developing water from formations composed of fine material of uniform grain size. A gravel-wall well is actually a large diameter drilled well, except that coarse material is placed around the screen instead of using the naturally occurring materials.

The most common construction used is the gravel envelope (Figure 2-5), where the outside casing is first sunk to the bottom of the formation. The diameter of this outer casing may vary from 18–72 in. (460–1,800 mm). Artificial gravel packs between 6- and 12-in. (150- and 472-mm) thick perform best and require the least maintenance. For example, a 24-in. (600-mm) bore would be finished with a 10- or 12-in. (250- or 300-mm) diameter screen and a 6–7-in. (150–180-mm) thick pack.

After the outer casing is in place, the screen is lowered to the bottom of the well and centered. Selected gravel is added to the annular space between the screen and the casing through a small diameter tremie pipe. The gravel is placed evenly around the screen in 2–4-ft (0.6–1.2-m) layers. As the gravel is added, the casing and tremie are slowly raised. The procedure continues until the entire screen is surrounded with gravel and the pack extends several feet (½–1 m) above the top of the screen. The outer casing is pulled back high enough to expose the entire screen section. As a rule, the screen is attached to an inner casing, extending to the land surface, into which the pump is placed. About 25 ft (8 m) of the outer casing is typically required to provide a seal against contamination by surface water. If the entire casing is removed, the gravel treatment must not extend up to or close to the land surface. The annular space between the working casing and undisturbed earth must be sealed with cement grout or puddled clay to prevent contamination from seeping into the formation. After the gravel filter has been placed, a pipe is often installed in the finished pump base or foundation to allow additional filter materials to be added if the gravel filter settles because of normal pumping operations, well development processes, or well rehabilitation procedures.

* The uniformity coefficient is the 40 percent retained grain size in a sieve analysis of sediment divided by the 90 percent retained grain size.

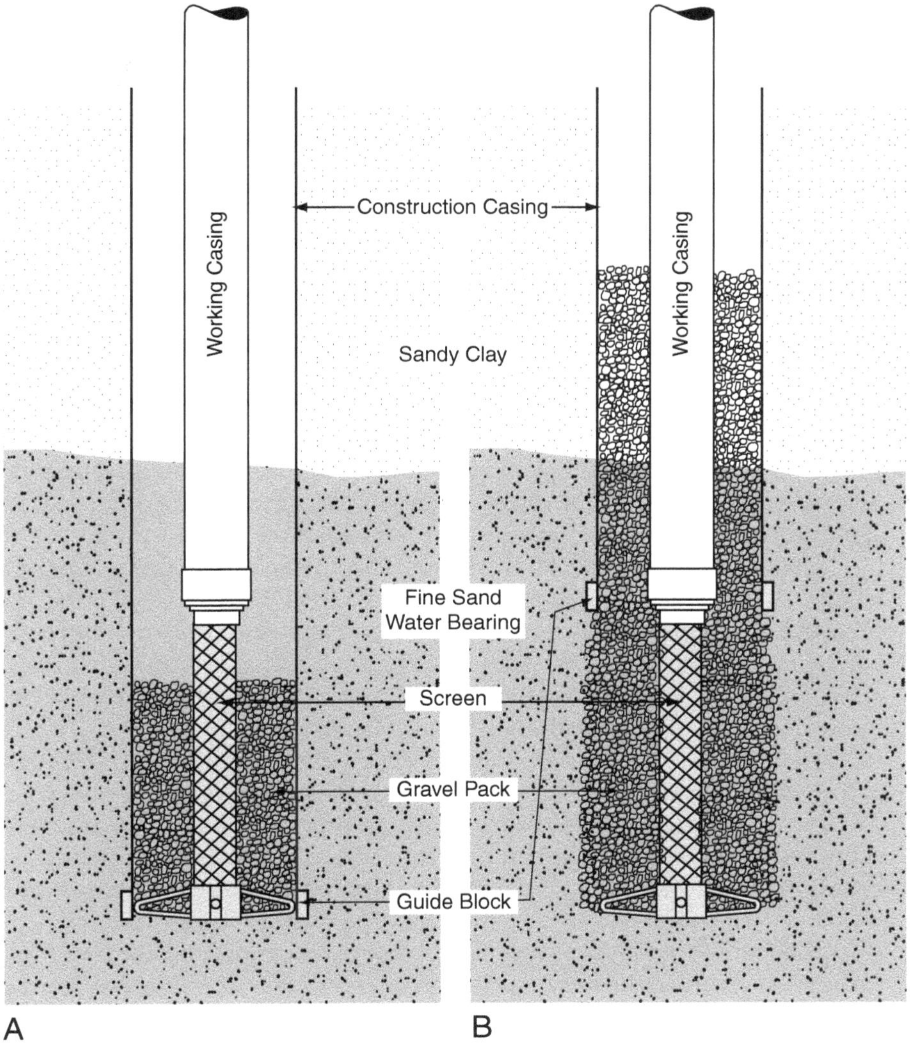

A—gravel-wall well with casing in place; B—completed gravel-wall well.

Figure 2-5 Two phases of gravel-wall well construction—gravel envelope method

COMMON WELL COMPONENTS

Well components common to most wells include well casings, cementing or grouting of wells, and well screens. These components, as they relate to well construction, are discussed in the following paragraphs.

Well Casings

The well casing is a lining for the drilled hole that maintains the open hole from the land surface to the water-bearing formation. Casings seal out contaminated water from the land surface and undesirable water from formations above the aquifer. For the casing to be entirely effective, it must be constructed of suitable materials and properly installed so as to be watertight for its entire depth.

Materials. Materials commonly used for well casings are alloyed or unalloyed steel, fiberglass, and polyvinyl chloride (PVC). Fiberglass and PVC have been used extensively in recent years for installations in shallow wells or where corrosion/bacteria may be of issue. In selecting a suitable material, the strain that the casing experiences during installation and the corrosiveness of the water and soil must be considered. All provide satisfactory service given the correct groundwater environment.

Many grades of casings are available, and ordering a nominal diameter of casing is not sufficient. The desired weight per foot of pipe or standard dimension ratio (SDR) of plastic casing must be specified. The tables in ANSI/AWWA A100-06 present data on steel pipe recommended for use as permanent well casings. ASTM F480 defines thermoplastic casing dimensions and joining. Bloetscher et al. (2022) and NGWA (1998) provide further detailed discussion on casing performance, selection, and installation, including the use of plastic casing types.

Lighter-weight materials may be used for test wells or temporary casings. Temporary casings are sometimes used as forms when a grout seal is placed around the outside of the permanent casing. The temporary casing is withdrawn as the grout seal is placed. Under such circumstances, lighter and less expensive material can be used.

Joints for permanent steel casings should have threaded couplings or should be welded to ensure watertightness from the bottom of the casing to a point above grade. This precaution will prevent surface contamination or undesirable underground waters above the water-bearing formation from entering the well. Thermoplastic casing is typically of either bell-and-socket construction, joined by cementing, or joined with spline-lock fittings in diameters relevant to water wells (ASTM F480).

Installation. When drilling a well by the cable-tool method, casing should be driven as soon as it becomes necessary to prevent the ground formation from caving. A drive shoe, attached to the lower end of the pipe, keeps the hole from collapsing. Drive shoes are threaded or machined to fit the pipe or casing, and the inside shoulder of the shoe butts against the end of the pipe. Drive shoes are forged of high-carbon steel, without welds, and are hardened at the cutting edge to withstand hard driving.

Casing is driven using drilling tools, drive clamps, and the drive head. A length of casing is attached to the previous length already set. A drive head is attached to the upper end of the casing to protect it from the driving blows of the drive clamp, which is attached to the drill stem. When the drill is lowered into the length of casing and subsequently raised and lowered, the action of the dropping clamp on the drive head forces the casing into the drill hole.

Wells constructed using rotary methods are not usually cased until drilling is completed. Because the casing is smaller than the drilled hole, no driving is required. In some instances, a casing is installed concurrently with drilling, such as with the use of dual-rotary or cable-tool drilling methods.

Special situations. If the formation being penetrated could likely cave equally throughout the full depth of the well, a single casing is usually sufficient. In these situations, the sand and gravel cave in around the outside of the casing and close the space between the drill hole and the casing.

If additional protection is desired against corrosion and pollution, an outer casing may be installed and the annular space between the casings filled with cement grout. With this type of installation, the outer casing may be either left in place or withdrawn completely. If withdrawn, the grout is placed as the temporary casing is removed. The temporary casing is generally one pipe size larger in diameter than the outside diameter of the couplings of the protective casing. This type of grouted installation may also be used where the water-bearing formation underlies clay, hardpan, or other stable formations.

Where the well penetrates water-bearing rock underlying unconsolidated material, the casing is driven into the rock to obtain a good seal. Unfortunately, a tight seal that will prevent pollution or the unconsolidated material above from entering the well is not guaranteed. One way to obtain additional protection is to drive the casing down to stable rock. The rock is then drilled to a depth of 10 ft (3 m) and under-reamed to a diameter that is 2-in. (50-mm) larger than the outside diameter of the shoe. The under-reamed portion of the drill hole below the bottom of the casing is filled with cement grout, and the casing is driven to the bottom of the hole. Before drilling resumes, the cement grout is allowed to set for several days, providing a good seal. Drilling then continues, and the cement grout is drilled out. An open, uncased hole is then constructed in the water-bearing rock below this point.

Fractured formations such as limestone that are channeled or creviced can yield polluted water or water of poorer water quality than desired. These formations should be protected unless overlain with an adequate thickness of unconsolidated formations or a competent layer of low-permeability rock. Under such circumstances, the well can be protected if it is watertight to a depth greater than that of the deepest existing well of questionable construction in the area and substantially below the lowest anticipated water level. The watertight construction is achieved by drilling the hole in the creviced rock that is 2-in. (50-mm) larger than the outside diameter of the casing couplings and filling the annular space between the drill hole and the outside of the casing with cement grout. In some areas, such construction may not be realistic because available water is cased off. Other methods of assuring adequate water quality protection may be necessary.

Cement Grouting

Water wells are cemented—or grouted—and sealed for the following reasons:

- to protect the water supply against pollution;
- to seal out water of an unsatisfactory quality;
- to increase the life of the well by protecting the casing against exterior corrosion;
- to stabilize soil or rock formations of a caving nature; and
- to prevent entry of stormwater runoff around the casing.

When a well is drilled, an annular space surrounding the casing is normally, and sometimes purposely, produced. Unless this space is sealed, a channel exists for the downward movement of water. In loose caving formations such as sand, the opening is usually self-sealing. In clay or other stable formations, this space must be cemented to prevent contamination from the land surface or creviced formations connecting with the surface.

When formations located below the depth of the protective casing are known to yield water of an unsatisfactory quality, such formations may be sealed off with liners set in cement grout for their entire length, which may be as much as several hundred feet deep. When a casing is extended to a consolidated formation lying below an unconsolidated formation, the best way to prevent sand or silt from entering the well at the bottom of the casing is by under-reaming and cementing. The casing exterior is protected against corrosion by encasing it in cement grout, as described earlier in the section on casing installation. A minimum grout thickness of 2 in. (50 mm) is recommended and may even be required by some regulatory agencies.

Materials. Materials used for cementing wells should facilitate proper placement and assume a permanent and durable form. Portland-cement grout, properly prepared and handled, meets these requirements adequately.

Proper preparation of the grout mixture is very important. The best results are obtained from neat cement and water mixed in the ratio of one bag of cement to not more than 5½ gal (20 L) of clean water. Under certain conditions, other materials may be used to accelerate or retard the time of setting, to lubricate the grout mixture, and to provide binders for sealing large crevices. A minimum of 2 percent, and a maximum of 5 percent, by weight of bentonite clay should be added to neat cement grout to compensate for shrinkage. Regardless of the materials used, cement, additives, and water must be mixed thoroughly.

Bentonite-Based Grout Mixtures

Bentonite grout mixtures (NGWA 1998) have several favorable characteristics, including the following.

- Bentonite, unlike cement-based mixtures, remains plastic when installed as a grout if it does not dry out and can be rehydrated if it does dry. High-active-solids bentonite seals do not crack or separate from surfaces. Low solids (or low-active solids) slurries will crack and separate in the vadose or unsaturated zone.
- Plastic, hydrated bentonite expands to fill voids, displacing air or water and other fluids.
- When properly prepared and emplaced, bentonite grout seals have hydraulic conductivities of 10^{-6} cm/s or less and reportedly as low as 10^{-12} cm/s. This wide range is a result of such variables as placement method (and skill), bentonite type used, conditions of the solids in the mixture, and the environment into which the seal is placed.
- Bentonite does not generate the heat of hydration experienced with cement, especially with larger annular radii (>2 in.).

Solids content and type. The amount of shrinkage is controlled by the bentonite grout's solids content; with bentonites that have high-reactive solids, content shrinking is far less than with low-solids types used in drilling fluid mixtures. Granular high-solids grades also have more dimensional stability than virtually liquid low-solids slurries (similar in some respects to the difference between neat cement and concrete). For these reasons, high-solids bentonites should be used instead of drilling mud bentonite for borehole-sealing applications.

Bentonite solids are more desirable than other solids. Unlike concrete, the preferable solids in bentonite are clays that provide bulk and dimensional stability (keeping the seal shape and size) in the bentonite gel matrix without increasing the permeability. Rock cuttings and sand have higher specific gravities than bentonite and tend to separate and sink in the hole. However, sand-dry bentonite mixtures (50:50) provide good stiff seals if placed so that separation is minimized (i.e., mixed in after the pump discharge and rapidly emplaced) and solids are permitted to hydrate downhole.

Bentonite grouts may shrink if moisture is lost. The presence of saline groundwater, strong acids or bases, or some organic compounds in contaminated groundwater may also cause desiccation and shrinkage. At times, a stiffer "set" is preferred than can be provided by bentonite alone, which is plastic and somewhat compressible. Cement is sometimes recommended to be added to a primarily bentonite mixture to provide this stiffer finished product. However, cement constituents serve to destroy the sealing properties of sodium bentonites. The trade-off for stiffness is generally in the form of a more brittle and more permeable seal. Calcium ions in the cement replace sodium ions by ion exchange, resulting in the clay particles settling closer together. On contact, the calcium ions link the platelets, causing flocculation. These changes are permanent once they are made.

In practice, major problems with cement addition to bentonite seals are the formation of cracks and failure to establish a seal with casing surfaces. This accounts for the generally higher permeability and may cause long paths of migration. For a stiffer, solid set, it is preferable to mix and place a very high-solids bentonite or bentonite-and-sand mixture instead of adding cement. Mixing water quality should be fresh (not saline) and approximately of drinking water quality in total dissolved solids (TDS) and calcium ion content. The water should be sanitary and free of foreign objects.

The grout must be applied in one continuous operation to ensure a satisfactory seal and be entirely in place before the initial set. The grout must always be introduced at the bottom of the space to be grouted to avoid segregation of materials, inclusion of foreign materials, or bridging of the grout mixture and, if above the fluid level, leaving large packets of air in the annulus.

Methods. Various methods are used for placing grout, including the dump-bailer method, air or water pressure drive, and pumping. Other proprietary methods of grouting, not discussed here, are used by well cementing companies.

The dump-bailer method is perhaps the simplest. The cement grout is lowered in a dump bailer that discharges its load when it reaches the bottom of the hole. After the necessary amount of grout is placed in the well, the casing is pulled up far enough so that the shoe is above the grout. A plug is placed in the bottom of the casing, which is then driven to the bottom of the hole, displacing the grout into the annular space around the outside of the casing.

If the annular space outside the casing is of sufficient width (1½–2 in. [40–50 mm]) to accommodate a grout (tremie) pipe, air or water-pressure drive is quite satisfactory. The tremie pipe should extend to the bottom of the annular space. Grout is then pumped into the pipe. The pipe should remain submerged in grout during the entire time the grout is being placed. The pipe may be left in place or gradually removed.

Grout pumping begins with installation of a pipe within the casing. The casing is suspended slightly above the bottom of the drill hole, and a suitable packer connection is provided at the bottom of the casing. The packer allows removal of the grout pipe and prevents grout leakage into the interior of the casing. Grout is pumped into the pipe and forced upward into the annular space. When the space is filled, the grout pipe is removed. Work on the well is not resumed for at least 72 hours, after which time the packer connection and plug are drilled out.

Screens

Generally, wells completed in unconsolidated formations such as sands and gravels are equipped with screens. Screens allow the maximum amount of water from the aquifer to enter the well with a minimum of resistance and prevent sand entering the well during pumping. Screens are sometimes installed in fractured formations that may collapse into the borehole and trap equipment in the borehole.

Although a screen prevents sand from entering the well during pumping, a screen also allows fine formation particles to enter the well during the development process so they may be removed by bailing. At the same time, the large particles of sand are held back, forming a permeable, graded natural gravel screen around the well screen itself. In this way, the hydraulic conductivity of the water-bearing formation around the well screen is greatly increased, resulting in lower velocity head loss and higher capacity per foot of drawdown.

Selection. Proper screen selection is extremely important in the design of a well drawing on unconsolidated aquifers. Selection is often a complicated matter that demands a highly specialized knowledge of well construction and operation. Consulting a reliable screen manufacturer is advised. Most manufacturers maintain a screen selection service and will make mechanical analyses of samples and recommend the proper opening.

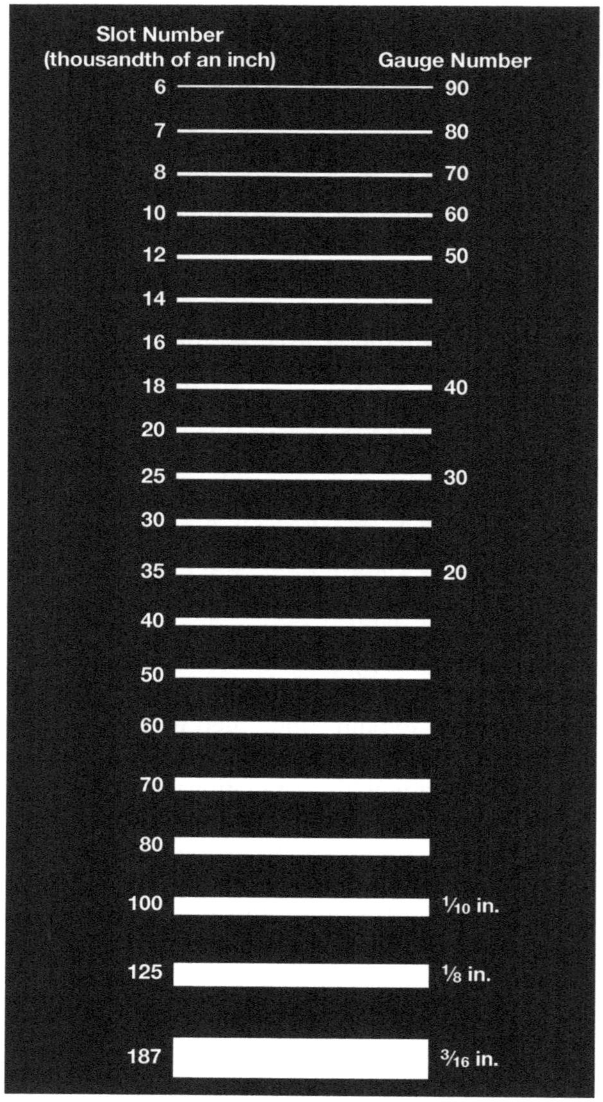

Figure 2-6 Screen slot size

Slot size. Depending on this type of well construction, the slot size is selected to permit a percentage of the formation material to pass through it. For naturally developed wells, this amount usually ranges between 35 percent and 65 percent, depending on uniformity of the material and the overlying formation. The size of screen openings, or the slot number, is usually expressed in thousandths of an inch as shown in Figure 2-6. The width of the slot, or slot size, is best determined with a mechanical sieve analysis of a sample of the water-bearing formation.

Representative samples of the formation must be selected for mechanical grain-size analyses. A complete descriptive log of the well should be submitted to the screen manufacturer with the samples. Information concerning the well diameter, aquifer thickness, transmissivity or hydraulic conductivity, and the desired well capacity should be included with the log.

Source: Courtesy of Frederick Bloetscher

Figure 2-7 Example of well screen with stabilizers

An artificial gravel-pack well may be constructed to allow an increase in screen slot size, as noted previously. When a well screen is surrounded by an artificial gravel wall, the size of the openings is controlled by the size of gravel used and by the types of openings as depicted in Figure 2-7. For select gravel-pack well construction, the passage of the pack material through the screen usually ranges between 10 and 30 percent, depending on the uniformity and gradation of the adjacent formations.

In the past, when casing slots were cut out with a knife or torch with very little open area, the water slot velocity was of concern. To promote better and more efficient well construction, the design criteria for water entrance through the screen opening between 0.1 and 0.2 ft/s (0.03 and 0.06 m/s) was adopted by many regulatory agencies' approval of construction. However, research and testing by D.E. Williams for the Roscoe Moss Company found that the actual head loss across just the thickness of a manufactured well screen, with the percentage of open area equal or greater than the specific yield of the aquifer, was insignificant until the flow-through velocity exceeded 2 ft/s (0.6 m/s). The very low screen velocity criteria promoted the use of large-diameter well screens and more efficient well construction. Of greater importance is determining the degree of turbulent flow that will be generated in the water flow through the formation and gravel pack material surrounding the well screen.

WELL DESIGN PROCEDURE

Various regulatory agencies have specific requirements for well permitting, design, construction, and approval. It is important to review and understand applicable requirements before beginning the process. The well design procedure starts with a review of the regulatory requirements and construction, along with some expectations on the potential water yield (and quality) from an aquifer. With many uncertainties, it is best to construct a small-diameter test well to ascertain the depth and thickness of the aquifer, conduct a pumping test and analyze the data to determine the transmissivity and hydraulic conductivity, and determine the water quality to be satisfactory or treatment needs. Typically, the client has a need for a specific water yield, and the well designer must bring these inputs together to design a cost-effective and efficient well.

It is often convenient to start with the well screen diameter, which should be at least one pipe size larger than the largest diameter of the pumping equipment to be installed. If a shroud needs to be installed around a submersible pump and motor, then an appropriate allowance in diameter needs to be made. If additional equipment is to be installed, such as a transducer or water-level controls, then an increase of two pipe size diameters may be needed.

If gravel-pack construction is to be used, then the borehole should meet minimum thickness requirements of 4 in. (100 mm) as specified in ANSI/AWWA A100-20. The purpose of Eq 2-1 to Eq 2-3 is to help select an appropriate borehole diameter for the well construction procedure to be used. If the average hydraulic conductivity, length of well screen and minimum borehole diameter, and the limit of borehole diameter are input, the limit of laminar flow can quickly be calculated. This value may appear to be very low, but most high-capacity wells are operated in the turbulent flow range. Typically, a flow yield of approximately four to six times the laminar flow rate will be cost-effective. The gravel-pack thickness can be increased to the available yield. Unfortunately, in low-permeability aquifers, the maximum practical well borehole diameter will limit the water yield. Other limitations such as saturated thickness, available drawdown, and static water-level depths affect the available yield.

To aid in the well design procedure, Nuzman (1989) developed some rule-of-thumb ratios between transmissivity and well specific capacity:

Confined aquifer $$Q/s = T/2{,}200 \quad \text{(Eq 2-1)}$$

Semi-confined aquifer $$Q/s = T/1{,}700 \quad \text{(Eq 2-2)}$$

Unconfined aquifer $$Q/s = T/1{,}200 \quad \text{(Eq 2-3)}$$

These ratios were developed for a typical well radius of influence of ½ mi and effective well diameter of 24 in. and assuming a storativity coefficient typical for the aquifer characteristics defined and the general assumptions of a theoretical aquifer (homogenous, isotropic, instant release from storage, infinite areal extent, and no leakage or recharge).

PUMPS

Pumps produce flow by transforming mechanical energy to hydraulic energy. Pump designs and applications are numerous, and energy specifications and ratings for pumps range from less than one to thousands of horsepower. To understand pumps and how they work, understanding the basic terminology is helpful.

Definitions

Capacity. The rate of flow delivered by a pump, in units such as gallons per minute, cubic feet per second, or barrels per hour. To calculate the power needed or the size of prime mover required to produce a desired capacity, the rate of flow and total dynamic head must be determined.

Dynamic head. The resistance to flow produced by a system. Dynamic head is equal to the sum of static head, velocity head, and friction head.

Static head. The static suction head plus the static discharge head (Figure 2-8). To calculate static head, all measurements in pumping are vertical, and the maximum drawdown is used as a reference. Measurements above this level are positive, and those below are negative. The same measuring procedure can be used for both submersible and surface-mounted pumps. Total static head is the combination of these two figures.

Table 2-1 Velocity-head data

Velocity (V) *fps*	Velocity Head (h_v)* *ft*	Velocity (V) *fps*	Velocity Head (h_v)* *ft*
1	0.02	11	1.87
2	0.06	12	2.24
3	0.14	13	2.62
4	0.25	14	3.04
5	0.36	15	3.49
6	0.56	16	3.97
7	0.76	17	4.44
8	1.00	18	5.03
9	1.25	19	5.61
10	1.55	20	6.21

* $h_v = V^2/2g$; g = acceleration due to gravity.

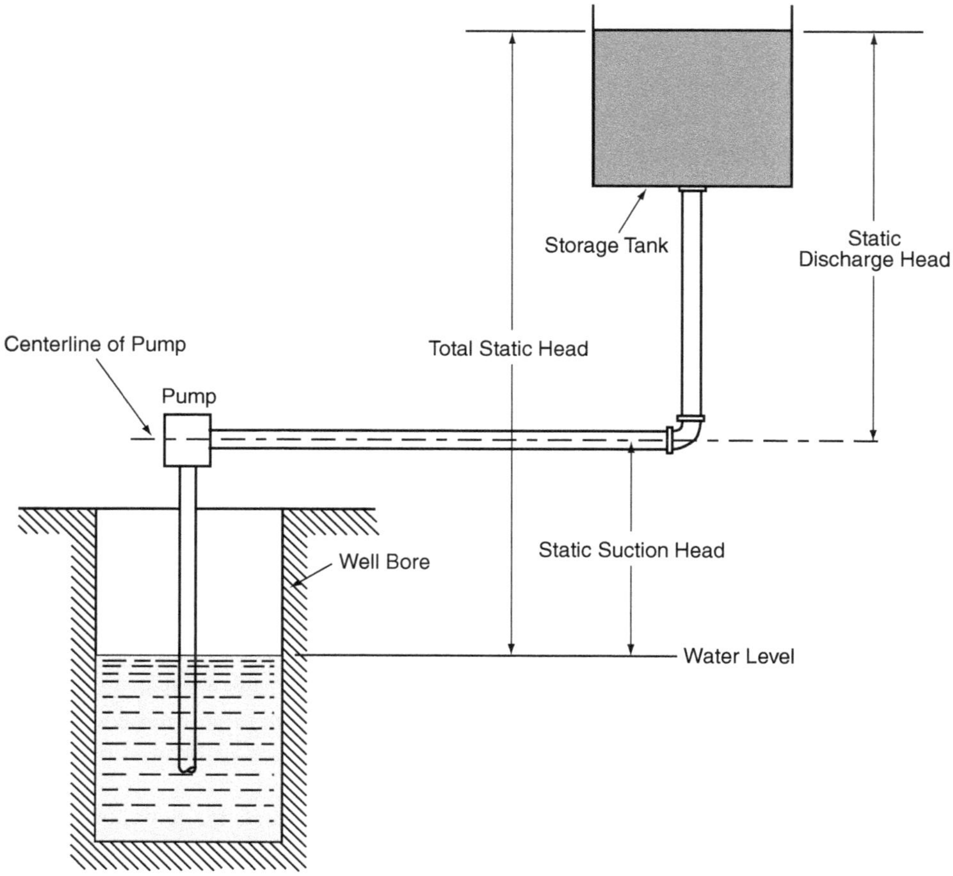

Figure 2-8 Schematic illustrating total static head

Static suction head. Vertical measurement, in feet, of the distance from the water level in a well to the pump centerline.

Static discharge head. The distance measured vertically from the pump centerline to the water level in storage.

Velocity head. The height through which a buoy must fall freely to attain its velocity. In most cases, the velocity head is small and can be ignored. Table 2-1 provides a way to determine velocity head.

Friction head. The loss of energy due to fluid motion along the inner surfaces of pipe and through fittings. With no change in elevation, friction head is the amount of head necessary to push fluid through pipe and fittings at the required velocity. Table 2-2 can be used to determine friction head when various pipe sizes and different flow rates are used. Friction head loss through fittings must be included (Table 2-3). Head losses for fittings are expressed in equivalent feet of pipe. For example, the loss through a regular 4-in. 90° elbow is equivalent to the loss through 13 ft of 4-in. pipe at the measured flow rate.

Table 2-2 Friction loss for water in ft per 100 ft (Schedule 40 Steel Pipe)

	2 in.		2½ in.		3 in.		4 in.	
Flow *gpm*	V *ft/s*	h_f Friction	V *ft/s*	h_f Friction	V *ft/s*	h_f Friction	V *ft/s*	h_f Friction
25	2.39	1.29						
30	2.87	1.82						
35	3.35	2.42	2.35	1.00				
40	3.82	3.10	2.68	1.28				
45	4.30	3.85	3.02	1.60				
50	4.78	4.67	3.35	1.94	2.17	0.662		
60	5.74	6.59	4.02	2.72	2.60	0.924		
70	6.69	8.86	4.69	3.63	3.04	1.22		
80	7.65	11.4	5.36	4.66	3.47	1.57		
90	8.60	14.2	6.03	5.82	3.91	1.96		
100	9.56	17.4	6.70	7.11	4.34	2.39	2.52	0.624
120	11.5	24.7	8.04	10.0	5.21	3.37	3.02	0.877
140	13.4	33.2	9.38	13.5	6.08	4.51	3.53	1.17
160	15.3	43.0	10.7	17.4	6.94	5.81	4.03	1.49
180			12.1	21.9	7.81	7.28	4.54	1.86
200			13.4	2.67	8.68	8.90	5.04	2.27
220			14.7	32.2	9.55	10.7	5.54	2.72
240			16.1	38.1	10.4	12.6	6.05	3.21
260					11.3	14.7	6.55	3.74
280					12.2	16.9	7.06	4.30
300					13.0	19.2	7.56	4.89
350					15.2	26.1	8.82	6.55
400							10.10	8.47
450							11.4	10.65
500							126	13.0
550							13.9	15.7
600							15.1	18.6

NOTE: The table shows average values of pipe friction for new pipe. For commercial installations, it is recommended that 15 percent be added to these values because no allowance for aging of pipe is included.

NOTE: Tables shown for Schedule 40 steel pipe are provided for example purposes only. For friction loss tables for other materials, consult the manufacturer.

(continued)

Table 2-2 Friction loss for water in ft per 100 ft (Schedule 40 Steel Pipe) *(Continued)*

	5 in.		6 in.		8 in.	
Flow *gpm*	*V* *ft/s*	h_f Friction	*V* *ft/s*	h_f Friction	*V* *ft/s*	h_f Friction
160	2.57	0.487				
180	2.89	0.606				
200	3.21	0.736				
220	3.53	0.879	2.44	0.357		
240	3.85	1.035	2.66	0.419		
260	4.17	1.20	2.89	0.487		
300	4.81	1.58	3.33	0.637		
350	5.61	2.11	3.89	0.851		
400	6.41	2.72	4.44	1.09	2.57	0.279
450	7.22	3.41	5.00	1.36	2.89	0.348
500	8.02	4.16	5.55	1.66	3.21	0.424
600	9.62	5.88	6.66	2.34	3.85	0.597
700	11.2	7.93	7.77	3.13	4.49	0.797
800	12.8	10.22	8.88	4.03	5.13	1.02
900	14.1	12.9	9.99	5.05	5.77	1.27
1,000	16.0	15.8	11.1	6.17	6.41	1.56
1,100			12.2	7.41	7.05	1.87
1,200			13.2	8.76	7.70	2.20
1,300			14.4	10.2	8.34	2.56
1,400			15.5	11.8	8.98	2.95
1,500					9.62	3.37
1,600					10.3	3.82
1,700					10.9	4.29
1,800					11.5	4.79
1,900					12.2	5.31
2,000					12.8	5.86
2,100					13.5	6.43
2,200					14.1	7.02

NOTE: The table shows average values of pipe friction for new pipe. For commercial installations, it is recommended that 15 percent be added to these values because no allowance for aging of pipe is included.

(continued)

Table 2-2 Friction loss for water in ft per 100 ft (Schedule 40 Steel Pipe) *(Continued)*

	10 in.		12 in.		14 in.	
Flow *gpm*	*V* *ft/s*	h_f Friction	*V* *ft/s*	h_f Friction	*V* *ft/s*	h_f Friction
650	2.64	0.224				
700	2.85	0.256				
750	3.05	0.294				
800	3.25	0.328				
850	3.46	0.368				
900	3.66	0.410	2.58	0.173		
950	3.87	0.455	2.72	0.191		
1,000	4.07	0.500	2.87	0.210	2.37	0.131
1,100	4.48	0.600	3.15	0.251	2.61	0.157
1,200	4.88	0.703	3.44	0.296	2.85	0.185
1,300	5.29	0.818	3.73	0.344	3.08	0.215
1,400	5.70	0.940	4.01	0.395	3.32	0.217
1,500	6.10	1.07	4.30	0.450	3.56	0.281
1,600	6.51	1.21	4.59	0.509	3.79	0.317
1,700	6.92	1.36	4.87	0.572	4.03	0.355
1,800	7.32	1.52	5.16	0.636	4.27	0.395
1,900	7.73	1.68	5.45	0.704	4.50	0.438
2,000	8.14	1.86	5.73	0.776	4.74	0.483
2,500	10.2	2.86	7.17	1.187	5.93	0.738
3,000	12.2	4.06	8.60	1.68	7.11	1.04
3,500	14.2	5.46	10.0	2.25	8.30	1.40
4,000	16.3	7.07	11.5	2.92	9.48	1.81
4,500			12.9		10.7	2.27
5,000			14.3		11.9	2.78
6,000			17.2		14.2	3.95
7,000					16.6	5.32
8,000						

Note: The table shows average values of pipe friction for new pipe. For commercial installations, it is recommended that 15 percent be added to these values because no allowance for aging of pipe is included.

(continued)

Table 2-2 Friction loss for water in ft per 100 ft (Schedule 40 Steel Pipe) *(Continued)*

	16 in.		18 in.		20 in.		24 in.	
Flow *gpm*	*V* *ft/s*	h_f Friction	*V* *ft/s*	h_f Friction	*V* *ft/s*	h_f Friction	*V* *ft/s*	h_f Friction
1,400	2.54	0.127						
1,600	2.90	0.136						
1,700	3.09	0.183						
1,800	3.27	0.203	2.58	0.114				
1,900	3.45	0.225	2.73	0.126				
2,000	3.63	0.248	2.87	0.139	2.31	0.0812		
2,500	4.51	0.377	3.59	0.211	2.89	0.123		
3,000	5.45	0.535	4.30	0.297	3.46	0.174	2.39	0.070
3,500	6.35	0.718	5.02	0.397	4.04	0.232	2.79	0.093
4,000	7.26	0.921	5.74	0.511	4.62	0.298	3.19	0.120
4,500	8.17	1.15	6.45	0.639	5.19	0.372	3.59	0.149
5,000	9.08	1.41	7.17	0.781	5.77	0.455	3.99	0.181
6,000	10.9	2.01	8.61	1.00	6.92	0.645	4.79	0.257
7,000	12.7	2.69	10.0	1.49	8.08	0.862	5.59	0.343
8,000	14.5	3.49	11.5	1.93	9.23	1.14	6.38	0.441
9,000	16.3	4.38	12.9	2.42	10.39	1.39	7.18	0.551
10,000			14.3	2.97	11.5	1.70	7.98	0.671
11,000			15.8	3.57	12.7	2.05	8.78	0.810
12,000					13.8	2.44	9.58	0.959
13,000					15.0	2.86	10.4	1.42
14,000					16.2	3.29	11.2	1.29
15,000							12	1.48
16,000							12	1.67
17,000							13.6	1.88
18,000							14.4	2.10
19,000							15.2	2.33

Note: The table shows average values of pipe friction for new pipe. For commercial installations, it is recommended that 15 percent be added to these values because no allowance for aging of pipe is included.

Table 2-3 Equivalent length of new straight pipe for valves and fittings for turbulent flow only

	Fittings		Pipe size, *in.*																				
			¼	⅜	½	¾	1	1¼	1½	2	2½	3	4	5	6	8	10	12	14	16	18	20	24
Regular 90° Ell	Screwed	Steel	2.3	3.1	3.6	4.4	5.2	6.6	7.4	8.5	9.3	11	13	--	--	--	--	--	--	--	--	--	--
		C.I.	--	--	--	--	--	--	--	--	--	9.0	11	--	--	--	--	--	--	--	--	--	--
	Flanged	Steel	---	--	.92	1.2	1.6	2.1	2.4	3.1	3.6	4.4	5.9	7.3	8.9	12	14	17	18	21	23	25	30
		C.I.	--	--	--	--	--	--	--	--	--	3.6	4.8	--	7.2	9.8	12	15	17	19	22	24	28
Long Radius 90° Ell	Screwed	Steel	1.5	2.0	2.2	2.3	2.7	3.2	3.4	3.6	3.6	4.0	4.6	--	--	--	--	--	--	--	--	--	--
		C.I.	—	--	--	--	--	--	--	--	--	3.3	3.7	--	--	--	--	--	--	--	--	--	--
	Flanged	Steel	--	--	1.1	1.3	1.6	2.0	2.3	2.7	2.9	3.4	4.2	5.0	5.7	7.0	8.0	9.0	9.4	10	11	12	14
		C.I.	--	--	--	--	--	--	--	--	--	2.8	3.4	--	4.7	5.7	6.8	7.8	8.6	9.6	11	11	13
Regular 45° Ell	Screwed	Steel	.34	.52	.71	.92	1.3	.7	2.1	2.7	3.25	4.0	5.5	--	--	--	--	--	--	--	--	--	--
		C.I.	--	--	--	--	--	--	--	--	--	3.3	4.5	--	--	--	--	--	--	--	--	--	--
	Flanged	Steel	--	--	.45	.59	.81	1.1	1.3	1.7	2.0	2.6	3.5	4.5	5.6	7.7	9.0	11	13	15	16	18	22
		C.I.	--	--	--	--	--	--	--	--	--	2.1	2.9	--	4.5	6.3	8.1	9.7	12	13	15	17	20
Tee-Line Flow	Screwed	Steel	.79	1.2	1.7	2.4	3.2	4.6	5.6	7.7	9.3	12	17	--	--	--	--	--	--	--	--	--	--
		C.I.	--	--	--	--	--	--	--	--	--	9.9	14	--	--	--	--	--	--	--	--	--	--
	Flanged	Steel	--	--	.69	.82	1.0	1.3	1.5	1.8	1.9	2.2	2.8	3.3	3.8	4.7	5.2	6.0	6.4	7.2	7.6	8.2	9.6
		C.I.	--	--	--	--	--	--	--	--	--	1.9	2.2	--	3.1	3.9	4.6	5.2	5.9	6.5	7.2	7.7	8.8
Tee-Branch Flow	Screwed	Steel	2.4	3.5	4.2	5.3	6.6	8.7	9.9	12	13	17	21	--	--	--	--	--	--	--	--	--	--
		C.I.	--	--	--	--	--	--	--	--	--	14	17	--	--	--	--	--	--	--	--	--	--
	Flanged	Steel	--	--	2.0	2.6	3.3	4.4	5.2	6.6	7.5	9.4	12	15	18	24	30	34	37	43	47	52	62
		C.I.	--	--	--	--	--	--	--	--	--	7.7	10	--	15	20	25	30	35	39	44	49	57
180° Return Bend	Screwed	Steel	2.3	3.1	3.6	4.4	5.2	6.6	7.4	8.5	9.3	11	13	--	--	--	--	--	--	--	--	--	--
		C.I.	--	--	--	--	--	--	--	--	--	9.0	11	--	--	--	--	--	--	--	--	--	--
	Reg. Flanged	Steel	--	--	.92	1.2	1.6	2.1	2.4	3.1	3.6	44	5.9	7.3	8.9	12	14	17	18	21	23	25	30
		C.I.	--	--	--	--	--	--	--	--	--	3.6	4.8	--	7.2	9.8	12	15	17	19	22	24	28
	Long Rad. Flanged	Steel	--	--	1.1	1.3	1.6	2.0	2.3	2.7	2.9	3.4	4.2	5.0	5.7	7.0	8.0	9.0	9.4	10	11	12	14
		C.I.	--	--	--	--	--	--	--	--	--	2.8	3.4	--	4.7	5.7	6.8	7.8	8.6	9.6	11	11	13
Globe Valve	Screwed	Steel	21	22	22	24	29	37	42	54	62	79	110	--	--	--	--	--	--	--	--	--	--
		C.I.	--	--	--	--	--	--	--	--	--	65	86	--	--	--	--	--	--	--	--	--	--
	Flanged	Steel	--	--	38	40	45	54	59	70	77	94	120	150	190	260	310	390	--	--	--	--	--
		C.I.	--	--	--	--	--	--	--	--	--	77	99	--	150	210	270	330	--	--	--	--	--

(continued)

Table 2-3 Equivalent length of new straight pipe for valves and fittings for turbulent flow only *(Continued)*

Fittings			Pipe size, *in.*																				
			¼	⅜	½	¾	1	1¼	1½	2	2½	3	4	5	6	8	10	12	14	16	18	20	24
Gate Valve	Screwed	Steel	.32	.45	.56	.67	.84	1.1	1.2	1.5	1.7	1.9	2.5	--	--	--	--	--	--	--	--	--	--
		C.I.	--	--	--	--	--	--	--	--	--	1.6	2.0	--	--	--	--	--	--	--	--	--	--
	Flanged	Steel	--	--	--	--	--	--	--	2.6	2.7	2.8	2.9	3.1	3.2	3.2	3.2	3.2	3.2	3.2	3.2	3.2	3.2
		C.I.	--	--	--	--	--	--	--	--	--	2.3	2.4	--	2.6	2.7	2.8	2.9	2.9	3.0	3.0	3.0	3.0
Angle Valve	Screwed	Steel	12.8	15	15	15	17	18	18	18	18	18	18	--	--	--	--	--	--	--	--	--	--
		C.I.	--	--	--	--	--	--	--	--	--	15	15	--	--	--	--	--	--	--	--	--	--
	Flanged	Steel	--	--	15	15	17	18	18	21	22	28	38	50	63	90	120	140	160	190	210	240	300
		C.I.	--	--	--	--	--	--	--	--	--	23	31	--	52	74	98	120	150	170	200	230	280
Swing Check Valve	Screwed	Steel	7.2	7.3	8.0	8.8	11	13	15	19	22	27	38	--	--	--	--	--	--	--	--	--	--
		C.I.	--	--	--	--	--	--	--	--	--	22	31	--	--	--	--	--	--	--	--	--	--
	Flanged	Steel	--	--	3.8	5.3	7.2	10	12	17	21	27	38	50	63	90	120	140	--	--	--	--	--
		C.I.	--	--	--	--	--	--	--	--	--	22	31	--	52	74	98	120	--	--	--	--	--
	Screwed	Steel	.14	.18	.21	.24	.29	.36	.39	.45	.47	.53	.65	--	--	--	--	--	--	--	--	--	--
		C.I.	--	--	--	--	--	--	--	--	--	.44	.52	--	--	--	--	--	--	--	--	--	--
Coupling or Union	Bell-Mouth Inlet	Steel	.04	.07	.10	.13	.18	.26	.31	.43	.52	.67	.95	1.3	1.6	2.3	2.9	3.5	4.0	4.7	5.3	6.1	7.6
		C.I.	--	--	--	--	--	--	--	--	--	.55	.77	--	1.3	1.9	2.4	3.0	3.6	4.3	5.0	5.7	7.0
	Square-Mouth Inlet	Steel	.44	.68	.96	1.3	1.8	2.6	3.1	4.3	5.2	6.7	9.5	13	16	23	29	35	40	47	53	61	76
		C.I.	--	--	--	--	--	--	--	--	--	5.5	7.7	--	13	19	24	30	36	43	50	57	70
	Reentrant Pipe	Steel	.88	1.4	1.9	2.6	3.6	5.1	6.2	8.5	10	13	19	25	32	45	58	70	80	95	110	120	150
		C.I.	--	--	--	--	--	--	--	--	--	11	15		26	37	49	61	73	86	100	110	140
Sudden Enlargement			$h = \frac{(V_1 - V_2)^2}{2g}$ feet of liquid; if $V_2 = 0$ $h = \frac{V_1^2}{2g}$ feet of liquid																				

NOTE: Tables shown for steel pipe and cast-iron pipe are for example purposes only. For equivalent head loss through valves and fittings for other materials, consult the manufacturer.

Centrifugal Pumps

A centrifugal pump uses centrifugal force to move a liquid through a change in elevation or against a total dynamic head. The pump consists of a suction nozzle, an impeller eye, an impeller (rotating element), a volute or diffuser, and a discharge nozzle. As fluid is drawn through the suction nozzle to the impeller eye, rotation of the impeller gives the fluid a high-velocity radial motion. Centrifugal force throws fluid from the outer tips of the impeller into the volute or diffuser and into the discharge line.

In both volute and diffuser types of centrifugal pumps, velocity head and, consequently, pressure are developed entirely by centrifugal force. In the volute-type pump (Figure 2-9), the impeller discharges fluid into a gradually expanding case. The volute efficiently changes part of the velocity head of the fluid leaving the impeller to pressure head. In the diffuser-type pump (Figure 2-10), the impeller is surrounded by progressively expanding passages of stationary guide vanes. The diffuser pump does a more complete job of converting velocity head to pressure and, consequently, is more efficient than the volute type.

Source: Courtesy of Frederick Bloetscher

Figure 2-9 Volute-type centrifugal pumps

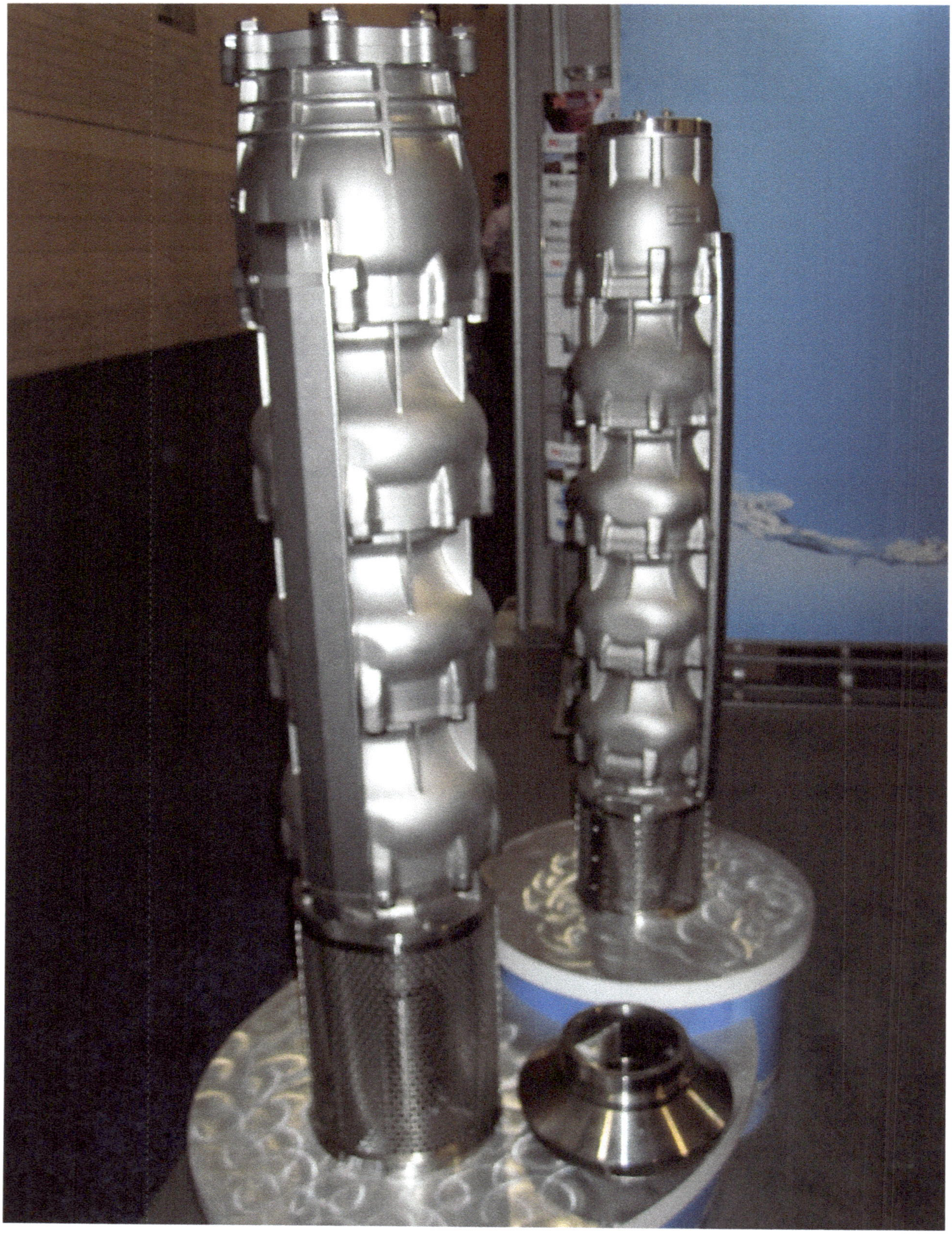

Source: Courtesy of Frederick Bloetscher

Figure 2-10 Multistage centrifugal pumps

The total suction lift or well depth below the pump centerline that can be pumped is regulated by the atmospheric pressure. If atmospheric pressure is 14.7 psi (6.9 kPa) and a perfect vacuum is present, this pressure could support a column of water equal in length to 14.7 × 2.307 = 33.9 ft (10.3 m). If the centrifugal pump could produce a perfect vacuum, the total theoretical lift would be 33.9 ft (10.3 m). Because a perfect vacuum at sea level is impossible to produce with a pump, the practical suction height varies from 60 to 85 percent of the theoretical possible distance depending on the efficiency of the installation. As altitude increases, the suction lift decreases. Because the use of a single-stage centrifugal pump is restricted to shallow wells (less than 20-ft [6-m] depths), other types of pumps are needed for deeper wells.

Deep-Well Turbine Pumps

In deep-well pumping, the reciprocating pump is superseded by the centrifugal pump as increased water volumes became necessary. Diminishing water tables, excessive costs of developing the deep pits that are used to place centrifugal pumps within reasonable suction lifts, and difficulty in providing efficient drivers fostered development of the deep-well turbine pump. The deep-well turbine pump is not truly a turbine but a combination of several stages of centrifugal impellers connected in series to a common shaft. The deep-well turbine pump, as illustrated in Figures 2-10 and 2-11, consists of the following:

- a prime mover (motor);
- a suitable shaft and bearings connecting the power source on the surface to impellers located under the well water;
- a series of impellers mounted in the bowl assembly at the lower end of the column that produces the required pressure head; and
- a discharge column pipe that channels water to the surface and acts as a housing and guide for the bearings and shaft assembly.

The deep-well turbine pump was designed for capacities as low as 10 or 15 gpm (40 or 60 L/min) and as high as 25,000 gpm (95,000 L/min) or more and for heads up to 1,000 ft (300 m). Most applications involve smaller capacities.

The pump illustrated in Figure 2-11 is a three-stage design. Each stage consists of a bowl, impeller, and diffuser manufactured as a standard unit. The number of bowls required for a particular installation depends on the dynamic head. The head determines the number of stages that must be provided. For large capacities, more than one pump may be needed. The capacity of the pumps used for bored wells is limited by the physical size of the well casing and by the rate at which water can be drawn without lowering its level to a point of insufficient pump submergence.

Submersible Pumps

A submersible pump is a turbine pump with its motors close-coupled beneath the bowls of the pumping unit (see Figure 2-12). The entire unit is installed underwater. This construction eliminates the need for the surface motor, long driveshaft, shaft bearings, and lubrication system of the conventional turbine pump. Submersible pump motors are cooled by water flowing vertically past the motor to the pump intake. The motor is usually longer and of smaller diameter than a surface motor of the same horsepower. When a large-capacity submersible pump is needed, the manufacturer should be consulted for specific design and installation recommendations.

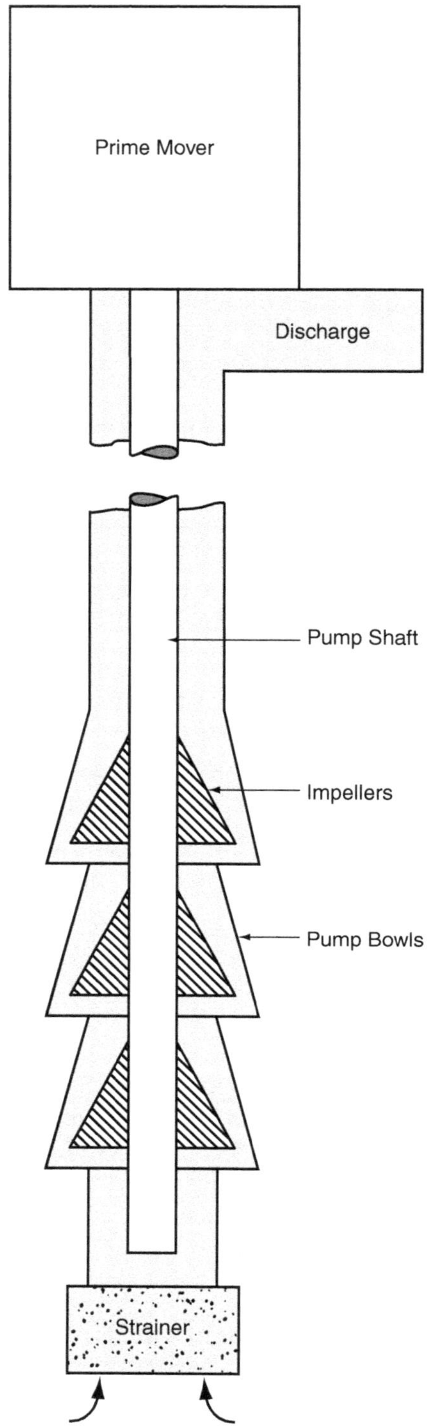

Figure 2-11 Vertical deep-well turbine pump (like Figure 2-10)

Source: Courtesy of Frederick Bloetscher

Figure 2-12 Submersible pump

The purchase and installation costs for a submersible pump may be higher or lower than for a conventional pump, depending on setting depth, required head and capacity, water corrosivity, and other factors. Operating costs may also be higher or lower based on motor efficiency, column bearing, hydraulic losses, cable losses, setting depth, and similar factors. A thorough analysis of all factors should be performed to compare surface and submersible motor-driven deep-well pumps for a specific installation.

Some inherent advantages of submersible pumps include:

- use in crooked well casings that are unsuited for other types of pumps;
- use in wells subject to flooding—the wells can be completely sealed;
- minimization of surface equipment; and
- silent operation.

Submersible pumps are especially useful for high-head, low-capacity applications such as domestic water supply. The pump should be at least 4 in. smaller than the casing pipe.

Propeller and Mixed-Flow Pumps

Propeller, axial-flow, mixed-flow, screw, and spiral-type pumps have found limited use in the production of shallow wells. These designs have open impellers, similar to a ship's propeller, and can be installed where flow is generally greater than 300 gpm (1,100 L/min) but where heads are less than 40 ft (10 m).

Rotary Pumps

A rotary pump combines the positive-discharge characteristics of the reciprocating pump with the constant, steady discharge of a centrifugal pump. Although a rotary pump uses a rotating element and appears similar to a centrifugal pump, positive displacement causes the pumping action. Specially designed runners squeeze the water between them as they rotate, building direct water pressure.

A well-designed rotary pump will create a relatively high vacuum, comparable to a centrifugal pump. However, rotary pumps are usually not as efficient as centrifugal pumps. Rotary pumps need to be well-designed and constructed of the best material or they will wear much faster. Still, the rotary-type pump is widely used.

Rotary-Displacement Pumps

A rotary-displacement pump (a positive displacement pump) is designed especially for relatively low capacities and for cased wells that are 4 in. and 6 in. (100 mm and 150 mm) or larger. Displacement of a piston in a cylinder of indefinite length causes the fluid flow. Figure 2-13 illustrates the pumping element, which consists of a main body made up of a stator and rotor, both of helical form, and the driveshaft assembly. The helices are worm threads; the stator has a double thread, and the rotor has a single thread.

As the rotor rolls on the inner surface of the stator, liquid is squeezed ahead by the rolling action, with minimum turbulence. The rotor is made of heat-treated stainless steel that has a hard, chrome surface to resist corrosion and abrasion. A one-piece bronze strainer with a rubber-seated foot valve keeps the column full of water, and no prelubrication is necessary. The stator is made of cutless rubber and is highly resistant to abrasion. Grit momentarily depressed into the rubber by the rotors is washed away by water when the rotor is released.

Reciprocating Pump

The oldest type of deep-well pump is the reciprocating or plunger-type pump (Figure 2-14). The reciprocating pump consists of a belt- or gear-driven head located above the highest water level in the well. A pulley drives a pinion shaft and, through suitable gearing, the plunger rod works up and down in the well. The prime mover is connected to the working, or pumping, barrel by pump rods.

The working barrel may be single or double acting. In the single-acting type, a check valve is located at the bottom of the cylinder, and a similar valve is in the plunger. The water flows into the working barrel through the check valve while the plunger is making its upstroke. On the downstroke, this water is held in the working barrel by the foot valve, and the plunger descends to the bottom of the barrel while the water passes through the valve in the plunger. On the next upstroke, the valve in the plunger closes, and the water above it is raised into the discharge pipe. At the same time, the foot valve opens, and the cylinder again fills with water.

In small-diameter wells, a check may be set in the casing below the water level and the plunger sized to the casing, which then becomes the working barrel cylinder. In this case, the rods work through a stuffing box at the top of the casing, and water is discharged out of a side-opening tee.

Double-acting cylinders discharge water on each downstroke and upstroke of the working head. Double-acting pumps can produce about 60-percent greater flow than pumps equipped with single-acting working barrels.

The capacity of this type of pump depends on the displacement of the liquid in the working barrel and the number of strokes per minute of operation. The pump is theoretically suitable for pumping wells of any depth, such depth being dictated by strength of material, power source, and economics.

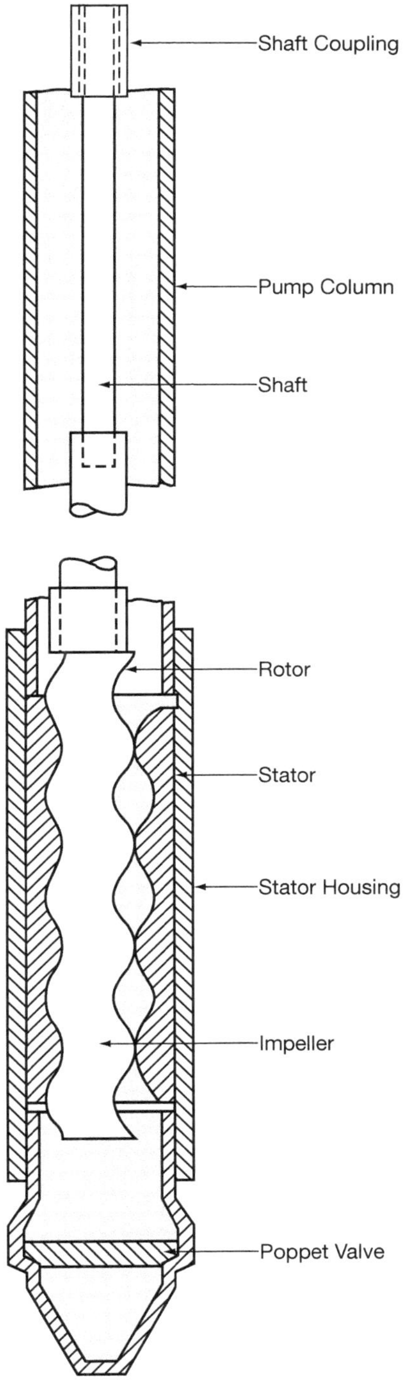

Figure 2-13 Rotary-displacement pump

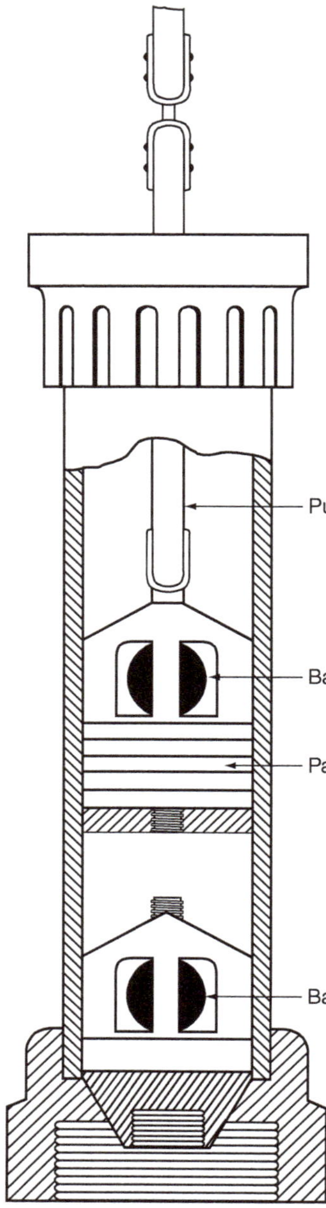

Figure 2-14 Plunger-type pump

Impulse Pumps

Impulse pumps produce pumping action by directly applying pressurized air or water. Airlift and jet impulse pumps are discussed in the following paragraphs.

Airlift pumps. Airlift pumps have capacities up to 2,000 gpm (8,000 L/min) and head to 1,000 ft (300 m). They are used in water wells, especially those containing sandy or corrosive fluids. The pump consists of a vertical pipe submerged in the well and an air-supply tube that feeds compressed air to the pipe at a considerable distance below the static water level. The mixture of air bubbles and liquid, lighter in weight than the liquid outside the pipe, rises in the pipe. A continuous flow of mixed water and air emerges at the top of the pipe, and new liquid from the well enters the pipe at the bottom (Figure 2-15).

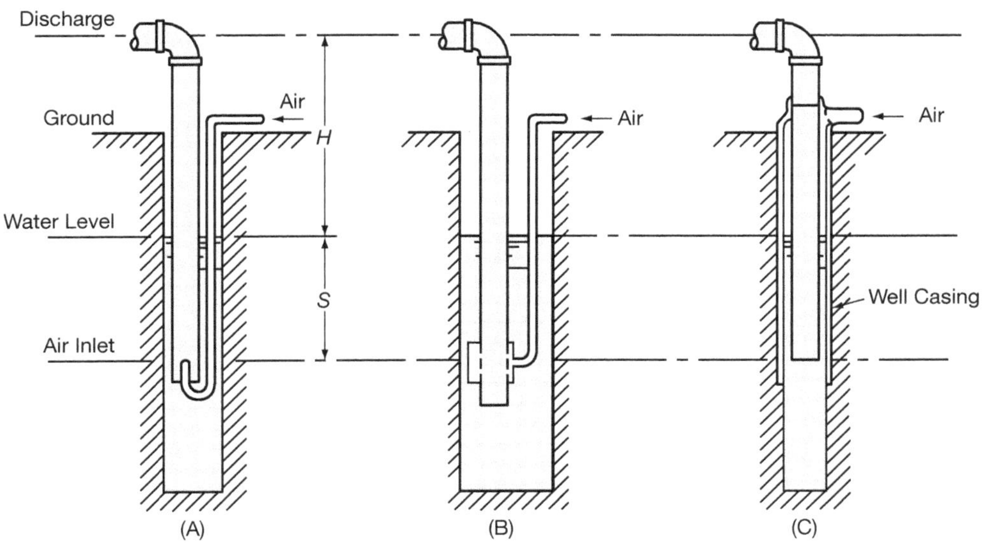

Figure 2-15 Airlift pump: (A) bottom inlet; (B) side inlet; (C) casing inlet

Because the only head-producing mechanism is the difference in specific weight of the water-air mixture inside the pipe and the water outside the pipe, the head that can be obtained from an airlift pump depends on the distance between water level in the casing and the elevation at which air is introduced. If head, *H*, is measured from the discharge pipe to water level, and submergence, *S*, is measured from water level to introduction of air, the ratio *H*/*S* is approximately one for most applications. The ratio can reach three for high heads (and low flows) and be as low as 0.4 for low heads (and high flows).

The volume of water pumped depends on the amount of air supplied. The pumping capacity increases with the amount of air supplied, up to an optimum. Because the discharge is a mixture of liquid and air, more air than optimum decreases the volume of water. The advantages of airlift pumps include:

- no moving parts;
- usability for corrosive and erosive fluids;
- gentle action (has been used to remove sand from buried undersea objects);
- operation on air (can be used in explosive atmospheres); and
- ability to be placed into wells of irregular shape where regular deep-well pumps cannot fit.

The disadvantages of airlift pumps include low efficiency (less than 40 percent) and the need for large submergences compared to conventional pumps.

Jet pumps. The jet impulse pump is shown in Figure 2-16. Water is forced down through a nozzle, forming a jet, and is discharged into the throat of a venturi diffuser at high velocity. The jet, or ejector, nozzle converts the pressure into velocity. Water discharges into the diffuser, causing a low-pressure area. Water then flows in from the well and mixes in the diffuser with the driving water. While passing through the tube, most of this high velocity is transformed into pressure, and this delivers both the driving water and the water drawn from the well to a high elevation.

The efficiency of this type of pump is rather low, about 25 to 30 percent, but the pump has no moving parts submerged in the well and is appealing where high capacities are not required. A jet pump is best used for a lift of 25 ft (8 m) or more and capacities less

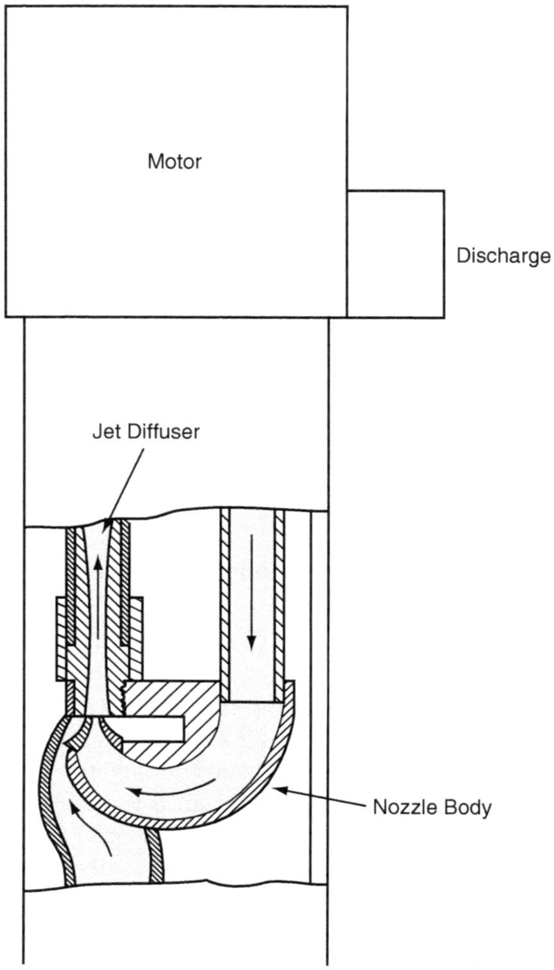

Figure 2-16 Jet-type deep-well pump

than 50 gpm (190 L/min). This pump is often used for 120–150-ft (40–50-m) lifts. In deeper wells, jet-pump efficiency becomes very low, and another type of pump should be used. Jet pumps are light and can pump very muddy or sand-loaded water. Centrifugal-jet pump combinations have been used to pump wells as deep as 400 ft (120 m); centrifugal pumps alone usually are limited to 20–25-ft (6–8-m) well depths.

PUMP SELECTION

Wells require a pump to get the water out of the formation and to the surface for treatment and use. As a result, the pump is the part of the well most connected to supply issues and therefore needs proper selection. Pump selection primarily depends on economics. The type of pump selected should give the best and most economical service over a prescribed number of years when pumping under specific conditions. Before selecting a pump, different types of pumping arrangements should be investigated. Conditions in the formation (e.g., microbial colonies) may drive pump materials (e.g., bronze is preferred where bacteria create havoc with steel and stainless steel). For any pump selection, the design engineer should tabulate the initial cost, cost of installation, cost of operation, cost of maintenance, and expected equipment life. The combination that offers minimum investment and operational costs and fulfills system requirements should be obvious from the calculations. In addition, operators need to monitor well performance against such conditions. If the pump performance deteriorates, it may indicate a host of issues, as discussed in Chapter 4.

Factors in Pump Selection

Some factors that must be considered before selecting a particular pump include:

- capacity,
- depth of well and pumping level,
- inside diameter of well,
- condition of borehole (straight or crooked),
- chemical and abrasive properties of the water (water quality),
- total head,
- type of available power,
- cost, and
- manufacturer-recommended maintenance and service schedule.

After a well has been tested for yield, operation requirements are dictated by demand. For some wells, a constant drawdown prevents production of sudden changes in water quality or quantity of sand. The size and condition of the well borehole are very important. For example, if a well is too crooked for efficient operation of a deep-well turbine pump and use of a plunger-type pump is not desirable, a submersible pump may be economical, even though initial installation might be more expensive. A chemical analysis will determine whether the pump needs to be built with materials to resist corrosion. Total head must be determined to lift or push the water to ground-surface elevation and to storage or delivery. A sufficient head needed to perform both tasks determines the pump horsepower required for accurate cost estimation.

Measuring pump performance. Energy cost is one of the principal expenses in pump operations. Pumps should be monitored to ensure that they are operating at or near peak efficiency. Total head, input horsepower, and quantity of water pumped must be measured.

The total dynamic head is the vertical distance from the water level in the well, while pumping, to the center of the free-flowing discharge, plus all losses between the point of entry of the water and the point of discharge. Losses in pipe can be obtained from Table 2-2, and pump-column losses are available from pump manufacturers' catalogs.

The water horsepower required to pump water can be determined by the following equation:

$$\text{whp} = \frac{\text{gmp} \times H_T}{3{,}960} \qquad \text{(Eq 2-4)}$$

Where:

gpm = the flow rate, in gallons per minute
HT = total head, in feet

The horsepower calculated using Eq 2-4 is for all equipment (pumps, prime mover, and the like) operating at 100 percent efficiency. Because 100 percent efficiency cannot be attained, the brake horsepower, or horsepower necessary at the pump shaft, is used. Brake horsepower can be obtained from manufacturers' data tables or performance curves. The pump efficiency, *Ep*, can be calculated from Eq 2-5.

$$E_p = \frac{\text{whp}}{\text{bhp}} \qquad \text{(Eq 2-5)}$$

The total horsepower required to operate the system is motor horsepower as follows:

$$\text{mhp} = \frac{\text{bhp}}{E_M} = \frac{\text{gpm} \times H_T}{3{,}960 \times E_P \times E_M} \qquad \text{(Eq 2-6)}$$

Where:

E_M = the efficiency of prime mover

The efficiency of an electric motor as a prime mover is usually between 60 and 95 percent, depending on size and type, but an exact value can be obtained from manufacturers' information.

The overall efficiency of a pump system depends on many factors such as specific speed, relative size, service materials, and physical characteristics of fluid. Large centrifugal pumps have developed more than 92 percent efficiency. The efficiency of smaller pumps may, in some instances, be only 20 or 25 percent.

To determine the overall efficiency of a pumping system, the efficiency of the pump, prime mover, and drive need to be included. The overall efficiency, E, then can be determined by the following equation:

$$E = E_p \times E_M \times E_D \qquad \text{(Eq 2-7)}$$

Where:

E_p = efficiency of the pump
E_M = efficiency of the prime mover
E_D = efficiency of the drive

$$\text{power required} = \frac{\text{theoretical power required}}{E} \qquad \text{(Eq 2-8)}$$

If an electric motor is used to drive the pump, the actual power required will be equal to the theoretical power times the result of Eq 2-8. To determine the cost of operation, one must convert horsepower to watts. One horsepower is equivalent to 746 W, and 1 kW is equal to 1,000 W; thus,

$$\text{kW} \cdot \text{h demand} = \frac{\text{gpm} \times H_T \times 0.746}{3{,}960 \times E} \qquad \text{(Eq 2-9)}$$

Power is usually sold in units of kilowatt-hours, and when Eq 2-9 is multiplied by number of hours used, to get:

$$\text{kW} \cdot \text{h} = \frac{\text{gpm}\ H_T \times 0.746 \times \text{hours}}{3{,}960 \times E} \qquad \text{(Eq 2-10)}$$

Total costs can be determined by multiplying Eq 2-10 by cost per kilowatt-hour.

$$\text{total cost} = \frac{\text{gpm} \times H_T \times 0.746 \times \text{hours} \times \text{cost/kW} \cdot \text{h}}{3{,}960 \times E} \qquad \text{(Eq 2-11)}$$

or

$$\text{power cost per hour} = \frac{\text{gpm} \times H_T \times 0.746 \times \text{cost/kW}\cdot\text{h}}{3{,}960 \times E} \qquad \text{(Eq 2-12)}$$

If a different type of power is used, cost per hour can similarly be calculated. Standby equipment should also be provided.

In general, pumps are driven by direct connection to prime movers or through right-angle drives. If the pump uses belts, it is old, and replacement should be considered. Electric motors and gasoline or diesel engines are usually prime movers for water-well pumps.

OPERATING CONDITIONS

Continuous operation of a pump is generally preferable to intermittent operation, but varying water demand usually requires some combination of off and on time. For improved well performance and pump life, system components and storage capacity should be designed to minimize the number of pump starts and stops per day. Although pump starts should be minimized, starting a pump several times per day or even more than once per hour adds only slightly to power consumption and normally gives acceptable life to pumps, motors, and controls. Variable speed drives can resolve this challenge. Variable speed drives can automatically ramp pump speeds up and down depending on demand, increasing pumping efficiency. The water hammer that is produced is the most destructive result of starts and stops.

For continuous running, a pressure-regulating valve or variable-speed drive is used that can match the pump output with the system demand. Such systems usually run at very low efficiency during low-demand periods. The overall cost of equipment and operation should be thoroughly analyzed before adopting such a system.

To achieve the lowest-cost operation, a system must run its pump and motor in the best efficiency ranges. Proper system components can assure this, but changing conditions sometimes justify altering or reselecting components to maintain economical operation.

For each type of pump and prime mover, the operating conditions must be checked against manufacturers' application information to ensure reliable operation. These operating conditions include the ambient air and water temperature ranges, pressures, flow, corrosive and abrasive factors, power supply variation, duty cycle, and protective devices.

Operational Limits of Pumping Units

A single-stage centrifugal-type pump offers satisfactory and economical service, and sizes are available for almost any desired capacity. However, application is restricted because the suction lift must not be more than about 22 ft (7 m), depending on elevation and temperature.

In a well that is clean and free of sand or grit, a rotary-type pump may perform as satisfactorily as a centrifugal pump, but rotary units are applicable only for operations that present low flow rates. Suction-lift specifications for rotary-type pumps are the same as for centrifugal pumps.

A centrifugal and jet, or ejector, combination pump can produce low rates of flow and suction lifts to 120 ft (37 m). In deeper wells, jets sometimes are used in combination with positive-displacement pumps.

For capacities exceeding a few gallons per minute (10 or 11 L/min) and settings deeper than 30 ft (9 m), a multiple-stage deep-well turbine pump that is driven directly by a submersible motor or through shafting by a surface-mounted motor or engine is usually selected. The choice between submersible and surface-driven turbines should be based on the following:

- analysis of initial costs and operating costs,
- maintenance/reliability,
- acceptability of aboveground structures and noise,
- likelihood of vandalism,
- well conditions,
- available power, and
- other factors specific to a particular installation.

Except for positive-displacement pumps, the discharge head increases as the rate of flow or capacity decreases, and the discharge head decreases as the rate of flow or capacity increases. If constant discharge under a varying head is to be maintained by a centrifugal pump, a variable-speed drive must be used. No problem is encountered when a positive-displacement pump is used because the capacity depends on the speed of the pump. The pressure that can be developed by a plunger-type pump is limited only by the size of the power unit and strength of materials.

ELECTRIC MOTOR SELECTION

Electric motors are usually selected according to National Electrical Manufacturers Association† standards, which include requirements for enclosures and cooling methods. An electrical specialist should be consulted for advice and assistance in selecting electric motors. AWWA's M2, *Instrumentation and Control*, also provides information on electric motors.

PUMP INSTALLATION

Proper pump installation increases pump efficiency, minimizes maintenance, and prolongs the life of piping. This section covers installation of pumps and associated piping.

Aboveground Installation

A good foundation, preferably concrete, should be constructed for pump placement. Foundation bolts should be placed according to the dimensions that are usually furnished by pump manufacturers. The pump must be easily accessible for regular inspection. Room should be provided for use of a crane, hoist, or tackle. Pits in which pumps are placed should be safeguarded against flooding.

Alignment. Pumps should be properly aligned by leveling the base with shims. Most pump bases, no matter how rugged, will spring and twist to some degree during shipment. Consequently, alignment is crucial when the unit is being installed.

Piping should line up naturally and be supported independently of the pump to eliminate strain on the pump casing; it should not be forced into place with flange bolts. The piping should be isolated from the pump head with a mechanical joint coupling near the head. After the piping has been installed, alignment should be rechecked. On unusually long discharge lines, a packed slip joint should be installed to compensate for elongation of pipe that might result from pressure or temperature changes.

† National Electrical Manufacturers Association, 2101 L St. N.W., Washington, DC 20037.

Piping. To protect the pump, a gate valve and check valve should be installed in the discharge pipe close to the pump. The check valve should be placed between the pump and a gate valve. If pipe connections are used on the discharge end of the pump to increase the size of discharge piping, the connections should be placed between a check valve and the pump. The selection of the discharge piping should be made with due reference to expected friction losses.

After the piping has been completed, alignment should be checked again using a straight edge and thickness gauge. The manufacturer's installation checklist and adjustment directions should be closely followed and double-checked before applying power to the pump unit. When pumping units have been aligned before piping is completed, piping strains that develop are probably the cause of any misalignments. Changes should be made accordingly. If stuffing boxes are adjusted properly and the pump and drives are aligned properly, the unit can usually be operated by hand with ease.

Deep-Well Installation

A deep-well pump driven by either a submersible motor or an aboveground driver must be installed according to the manufacturer's instructions. The pump must be sized and set so that it will never run for even a few minutes at no delivery, which could occur if excessive drawdown is present or if the pumping level is lowered to the intake area, because the groundwater cools the pump. Running with little or no delivery is likely to damage the pump bearings (if they run long enough to heat up and boil) and cause overheating failure of a submersible motor. If the well drawdown or the delivery system could cause the pump to run at little or no delivery, protection should be provided to ensure flow. Such protection could be in the form of a flow switch or well-level switch that would shut off the pump or sound an alarm if flow or water level dropped below a safe minimum. The minimum water level above the pump intake should always be kept greater than the required suction head (net positive suction head [NPSH]) specified by the manufacturer. The water level should not be drawn down into the vicinity of the pump. Pump capacities should be selected to ensure at least 5–10 ft (1.5–3 m) of water over the top of the screen at maximum drawdown. As discussed previously, cementing provides sanitary protection.

The materials used in the pump and delivery system must be resistant to significant corrosion caused by normal water conditions in the well or any periodic chemical cleaning operations performed with the pump in place. Additionally, the well must be properly designed and developed before installing the production pump to minimize sand pumping. Most pump and submersible-motor warranties do not cover failures from abrasive damage and corrosion.

Check valve. Unless some unusual requirement prevents it, a check valve should always be installed within 25 ft (8 m) of a deep-well surface-driven or submersible pump. The check valve prevents problems that may occur when water in the delivery pipe flows back into the well when the pump is turned off. These problems include:

- Backwashing, which can disturb the stabilized particles located outside screens and perforations, often increasing sand and turbidity in the well.
- Backflow, which may spin the pump at high speed in reverse, causing damage or shortened life. This problem will not occur if the pump is designed to withstand high speed or if it is equipped with a device to prevent backspin. Attempting to restart a pump during backspin decreases bearing life and may cause tripping of protective devices with prolonged starting current.
- Refilling of the delivery pipe at each start, which wastes power.
- Creation of a vacuum and water hammer. Aboveground check valves and shutoff valves near the pump, which are often required, can create a vacuum in a section

of the delivery pipe after the pump turns off. This occurs because atmospheric pressure can only support water in the pipe to less than 34 ft (10 m) above the level in the well. When this evacuated section refills on starting, the moving water striking the stationary water at the closed valve creates a severe hydraulic shock (water hammer), which can cause pipe, valve, pump, or motor failure. An air and vacuum release valve should be installed between the pump head and the check valve. Weighted check valves can reduce water hammer.

Additional check valves may be required, depending on setting depth, valve rating, and aboveground equipment. A check valve in the delivery pipe of a submersible pump will hold the pipe full of water if the pump is removed from the well. Special check valves are often used in which a small replaceable plug can be broken off to create a drain by dropping a weight down the well before pulling the pump.

The manufacturer's installation documentation for surface-driven, deep-well pumps includes information about the following:

- column pipe assembly,
- bearings and shafting,
- lubrication,
- alignment,
- mounting and aligning of the aboveground drive,
- setting of the impeller position, and
- use of proper controls.

Submersible pump information includes:

- settings that prevent motor burial in sand or silt,
- water temperature and flow past the motor to provide proper cooling,
- use of cable and splices that meet the amperage and voltage requirements,
- pipe tightening to prevent unscrewing by motor-starting torque,
- clamping of cable to delivery pipe,
- proper controls and protections, and
- necessary checks before, during, and after installation.

SANITARY PROTECTION

All water supply wells must be provided with adequate sanitary protection through proper construction and disinfection. This includes bentonite seals and protection against surface flooding.

Sanitary Seal Construction

Wells must be constructed in formations sufficiently deep to be protected from surface contamination. The minimum depth of safe water will vary with soil formations and surrounding conditions. In unconsolidated materials, water from depths of 25–30 ft (8–9 m) or more is reasonably protected. The well casing should extend at least to that depth, and the screen should be set below it.

If a well must be constructed at a depth less than that recommended in pervious material, an impervious layer of soil at the land surface can provide some protection. A layer of well-compacted clay at least 2-ft (0.6-m) deep should be placed on the land surface for a radius of 50 ft (15 m) around the well. The clay layer will minimize percolation from surface water and tend to divert it to the edge of the clay and away from the well. Every well casing should be grout-sealed from land surface to the full depth of the root zone (30–50 ft [9-15 m]) or standing water levels, whichever is less. Many regulatory agencies require a minimum of a 6 ft^2 (0.55 m^2) square concrete pad around the well casing. Typically, the pads are sloped 2–3% away from the well.

Well Development

As noted in Bloetscher et al. (2022), often the most overlooked part of well construction is development. The point of well development is to remove any mud, sand, silt, or other materials introduced into the well as a part of the drilling process. Mud removal caked on walls is a common challenge. Bloetscher et al. (2022, 2008) identified several of methods for well development, depending on the well type, aquifer type, grain size distribution, available equipment, well construction methods, and installed materials and equipment:

- overpumping,
- raw-hiding,
- pump surging,
- high-velocity jetting,
- air surging and pumping,
- mechanical surge blocks,
- double-flanged swabbing, and
- chemical additives.

Each method has its advantages and disadvantages, and each is descriptive. For example, overpumping simply means pumping at a higher rate than normal. Overpumping is the easiest method to use and may be satisfactory in many cases. Surging requires varying pump rates and direction, as does air-surging. The concept is to introduce a water hammer into the formation to loosen unwanted materials. Inserted swabs and blocks are designed to remove mud cake materials. A variety of chemicals can be used. Acidization is common to remove materials that cannot be swabbed or scoured off. Dispersing agents can also be used, but most chemical action is unnecessary for initial development.

Records of well development should be maintained. Establishing an efficient and timely protocol for development will ensure a smooth operation in the field. Such a protocol might be as follows (Bloetscher et al. 2005):

1. Measure and record flow.
2. Measure and record pumping distance to water level.
3. Turn pump off and wait 5 min.
4. Record static water level.
5. Calculate drawdown (pumping water level minus static level).
6. Calculate specific capacity (specific capacity is the ratio of each well's yield to its drawdown and is expressed in gallons per minute per foot of drawdown (gpm/ft)).
7. Develop the well by surging or air lifting.
8. Repeat steps 1–7.

If the well's specific capacity increases significantly (at least 25 percent), disinfect the well and place it back into service. Otherwise, keep repeating steps 1–7.

Any development effort, whether for plugging or not, will create a wastewater product that may contain chemicals, silt, sand, or other debris. The quality of this wastewater may require treatment. In addition, all well development has some risk of disturbing formation and increasing the potential for sand entering the well during operations/pumping. Hence, some care should be taken during well development.

Disinfection

During well construction, the drill hole is subject to contamination from the land surface. Contamination can also be introduced by tools, drilling mud (in the case of the rotary method), the casing, and the screen. Normally, extended pumping would rid the well of this contamination; however, disinfecting the well with chlorine is faster.

Many disinfection methods are available and should be selected by the engineer supervising the installation. Generally, sufficient chlorine must be thoroughly mixed with the water in the well casing to produce a concentration of at least 50 mg/L in the well when the disinfectant is pumped into the well. This solution must come in contact with the pump and discharge piping. Disinfection is achieved by adding chlorine in the casing and producing a mix by alternately starting and stopping the pump or by other methods. Contact time is a minimum of 24 hours.

The material used for gravel treatment, even though washed and clean, still carries contamination. Therefore, gravel-wall wells are sometimes difficult to disinfect following construction. In addition to the procedures outlined for disinfection, a tablet or powdered calcium hypochlorite can occasionally be added by hand to the gravel filling tube as the gravel is placed.

Even with disinfection, the water pumped from a well may still show evidence of contamination. Under such circumstances, a chlorinator can be installed at the well to treat all the water discharged to the system. In time (perhaps as long as three or four months), normal pumping will usually rid the well of contamination. During this period, a free chlorine residual will make it possible to use the water. Additional information on disinfection is available in ANSI/AWWA Standard A100-20.

REFERENCES

American National Standards Institute/American Water Works Association (ANSI/AWWA). 2006. *Standard A100-06.* Denver: AWWA.

ASTM. 2022. F480 Standard Specification for Thermoplastic Well Casing Pipe and Couplings Made in Standard Dimension Ratios (SDR), SCH 40 and SCH 80. West Conshohocken, PA: ASTM.

Australian Drilling Industry Training Committee. 1997. *Drilling: The Manual of Methods, Applications, and Management.* Boca Raton, Fla.: CRC Press/Lewis Publishers.

Bloetscher, F., A. Muniz, and G.M. Witt. 2005. *Groundwater Injection: Modeling, Risks, and Regulations.* New York: McGraw-Hill.

Bloetscher, F., A. Muniz, and J. Largey. 2008. *Siting, Drilling, and Constructing Wells.* Denver: AWWA.

Bloetscher, F., A. Muniz, and J. Largey. 2022 (2nd ed.). Water Supply Wells: *Siting, Drilling and Construction.* Denver: AWWA.

Driscoll, F.G. 1986. *Groundwater and Wells.* St. Paul, Minn.: Johnson Division.

National Groundwater Association. 2017. *Manual of Water Well Construction Processes*. Westerville, Ohio: National Ground-water Association.

Nuzman, C.E. 1989. "Well Hydraulic Flow Concept" In: *Recent Advances in Groundwater Hydrology*. Minneapolis, Minn.: American Institute of Hydrology.

Roscoe Moss Co. 1990. *Handbook of Groundwater Development*. New York: Wiley-InterScience.

Williams, D.E. 1985. "Modern Techniques in Well Design." *Journal AWWA*. 77(9):68–74.

This page intentionally blank.

Chapter 3

Operations: Tracking Data and Identifying Problems

As part of effective wellfield management, the proper operation of a groundwater system includes the gathering, compiling, and recording of a wide variety of data. The main objective of any recordkeeping program is to compile information that makes it possible to identify changes in operating characteristics and conditions over time. Such information can be used to guide improvements to the physical system and the operating methods. For example, pumping data from production wells can be used to detect a loss of production efficiency and, possibly, the cause of the loss. Data on the operating history of a well can also be used to schedule well maintenance at opportune times to prevent breakdowns. Similarly, water quality data collected under different hydrologic conditions and over the long term can help identify vulnerabilities to contamination. For example, seasonal fluctuations in the quality of water from deep public-supply wells are more likely to be related to seasonal differences in well operation than to changes in the quality of water in the aquifer itself.

It is critically important that records be collected and kept in the development, maintenance, and rehabilitation of wells. Although it is not important which format is used in recordkeeping, the date and time must be kept for each set of measurements. Other information that should be recorded is presented in this chapter. The information recommended for collection includes both design and construction data and operational data. It is also critically important that operators understand the requirements pertaining to wells in their state, which may be associated with the primary agency responsible for public water supply regulatory compliance or other water resource agencies. Specific logs and records may be required for regulatory compliance purposes, and utilities should be familiar with those requirements to maintain and/or provide adequate records.

DESIGN AND CONSTRUCTION RECORDS

The following outlines data to be included in the construction specifications or to be defined by the hydrogeologist during construction:

- well diameter,
- proposed total depth,
- position of the screens (or portion of the open hole if constructed in rock),
- method of construction and materials to be used for the screen and casing, and
- pump design.

The following should be obtained after the drilling process is complete and the casing and well screens have been put in place and stored in a reference place so that this information can be readily retrieved:

- water quality analyses (see Water Quality section in this chapter and Chapter 5),
- detailed individual well (geologic) logs,
- static (nonpumping) water levels in the aquifer, and
- design pump discharge pressures.

When the production well has been constructed, "as-built" records of the well should be recorded. These records should include:

- the method of construction used to drill the well,
- the driller's log of the materials encountered during drilling,
- geophysical logs,
- well development records,
- diameters (and materials of construction) of well casing and screens,
- slot sizes of the screen,
- the depths (settings) of the casing and screen,
- the aquifer in which the well was completed,
- surveyed location of completed well,
- surveyed elevation of water-level measurement point, and
- the total depth of the well.

PUMP DATA

Pump data should include:

- the type (and make) of the pump installed,
- the type and horsepower of the motor (driver),
- the pump setting (depth to the pump intake),
- the setting of the air line or other device for measuring the water level in the well,
- notation for the point (and reference elevation) used for measurement of the water level,
- all information provided by the pump and motor manufacturer, such as capacity and efficiency data, and

- maintenance information from manufacturer's literature; to maintain warranties, the manufacturer's recommended maintenance needs to be conducted.

WELL ACCEPTANCE AND PUMPING TEST

A newly constructed well with a new pump should be evaluated as a unit to establish a standard against which to measure performance in the future. Data should be collected to determine well losses that relate to well design and construction. Comparative data pertaining to the physical condition of the pump unit should also be collected. Proper study and comparison of such data enable the operator (or consultant) to anticipate maintenance and repair needs. Of course, similar tests should be rerun after any maintenance or repair work. This data should include:

- water-level measurements made before, during, and after the (drawdown) pumping test,
- a record of the pumping rate,
- hydrographs generated during the test,
- any raw data collected (manual or computer generated), and
- a copy of the hydrogeologist's report on the procedures and test results.

EVALUATING WELL PERFORMANCE

Monthly Pumpage

The total pumpage for each well should be recorded monthly and graphed to illustrate the seasonal and yearly production rates. In addition, the number of days each well was pumped during the month and the daily peak demand for the month should be recorded. These data can be used for future projection of water withdrawal rate and to monitor the actual volume of water produced from each well to predict periods between maintenance. These data can usually be recorded from a totalizer on flowmeters installed in the discharge piping for each well. The data show the utility how much water is pumped from each well, which is useful for other calculations for well performance. If there are withdrawal permits required for the well operation, this information will need to be submitted monthly to the regulatory agency.

Water Levels

Records of water levels in the well during periods of nonuse (static) and during pumping should be recorded. The static levels can identify changes in the amount of water that may be available in the aquifer with time. The levels also provide a baseline for determining the amount of drawdown. Measurements of water level in observation and monitoring wells can also be used to evaluate static water levels in the aquifer at any point. All wellfields should include monitoring wells for water quality and water elevation purposes. Monitoring pumping levels for any well will illustrate the seasonal variations of pumping rates, river stage, water temperatures (well water and river), and drawdown with time. Trends in the pumping level with time can reflect losses in efficiency in the well over time.

Water Temperature

The groundwater temperature should be recorded and plotted to obtain base data for determining future expectations of groundwater temperature. As the temperature of the groundwater varies, the capacity of the well fluctuates because of the viscosity of the water. Viscosity offers resistance to flow and is expressed as a coefficient of dynamic viscosity—or the force required to move a unit area a unit distance. In areas where recharge to the aquifer may come from infiltration of surface water, the temperature of the adjacent surface water body should also be recorded. The temperature records are used to evaluate infiltration, recharge, and where groundwater is under the influence of surface water.

Water Quality

It is desirable to collect water quality data that can be used to identify geochemical conditions in the surrounding aquifer. Geochemical conditions—especially redox conditions, pH, and alkalinity—strongly affect the mobility and persistence of many drinking-water contaminants in groundwater. More specifically, geochemical conditions influence whether a contaminant that has been released to the groundwater will travel with the groundwater, react with the aquifer material, or degrade before reaching the well.

Redox conditions affect the stability of nitrate, arsenic, iron, manganese, certain volatile organic compounds (VOCs), etc. Redox conditions (e.g., oxic and anoxic conditions) in an aquifer(s) surrounding the well can be assessed by analyzing a sample from the well for six water quality constituents that are relatively inexpensive and easy to measure—dissolved oxygen (O_2), nitrate (NO_3), manganese, iron, sulfate (SO_3), and hydrogen sulfide (H_2S)—and comparing the measured concentrations to threshold concentrations in the redox framework illustrated in Table 3-1. As an example, if water quality data for the well indicates the presence of iron- or sulfate-reducing conditions [Fe(III) or sulfate reduction], nitrate in the surrounding aquifer would be unlikely to reach the well and be detected in water from the well. Wells with open intervals that cross redox boundaries (identified as having mixed redox conditions) can be vulnerable to a wide range of chemical contaminants. In addition, redox boundaries provide an ideal habitat for bacteria that cause biofouling. Knowing that a redox boundary is present within the open interval of a well may provide a heads-up that the well might be a high-maintenance well.

Water quality data should be collected over different hydrologic conditions and over the long term. All water quality data should be routinely graphed to identify patterns in the quality of water from the well. Important changes in water quality over time are

Table 3-1 Framework for identifying redox conditions of groundwater from a well*

General redox category	Predominant redox process	Water-quality criteria (mg/L)					Fe/H_2S mass ratio
		O_2	NO_3-N	Mn	Fe	SO_4	
Oxic	O_2 reduction	≥0.5	--	<0.05	<0.1	--	--
Suboxic		<0.5	<0.5	<0.05	<0.1	--	--
Anoxic	NO_3 reduction	<0.5	≥0.5	<0.05	<0.1	--	--
	Mn(IV) reduction	<0.5	<0.5	≥0.05	<0.1	--	--
	Fe(III) or SO_4 reduction	<0.5	<0.5	--	≥0.1	≥0.5	--
	Fe(III) reduction	<0.5	<0.5	--	≥0.1	≥0.5	>10
	Mix Fe(III) and SO_4 reduction	<0.5	<0.5	--	≥0.1	≥0.5	≥0.3 and ≤10
	So_4 reduction	<0.5	<0.5	--	≥0.1	≥0.5	<0.3
	Methanogenesis	<0.5	<0.5	--	≥0.1	<0.5	--
Mixed		Criteria for more than one redox process are met					

Adapted from: Jurgens et al. 2009
* The redox framework is useful even if the input dataset does not include all six recommended constituents.

not limited to changes in concentrations of drinking water contaminants but can include changes in the general characteristics of the water, such as temperature, dissolved oxygen, pH, alkalinity, and dissolved-solids concentrations.

Temporal fluctuations in water quality can help identify the presence of preferential flow pathways, i.e., pathways that provide little resistance to flow, which can increase the vulnerability of the well to contamination. Short-term water quality fluctuations in deep public-supply wells should not be expected in the absence of preferential flow pathways. For example, a well in Modesto, Calif., had higher concentrations of nitrate, uranium, and alkalinity during winter months. This was because the well was functioning as a conduit for the downward movement of contaminants during times of light or no pumping due to a downward hydraulic gradient in the surrounding aquifer. These conditions allowed shallow contaminated water to accumulate at depth around the wellbore, only to be drawn back into the well when heavier pumping resumed (Figure 3-1). The quality of the water from the well improved once the pumping schedule was updated to include longer pumping periods during winter months.

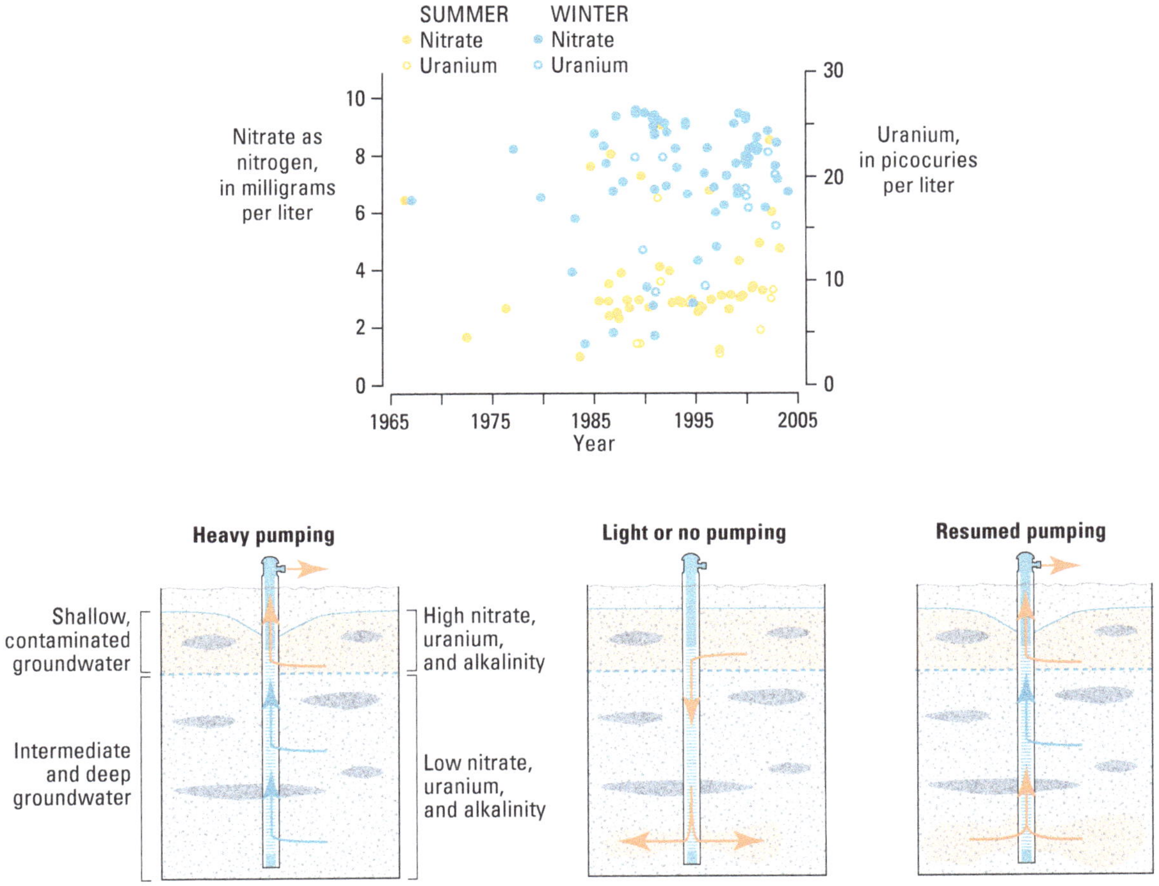

Source: Eberts et al. 2013

Figure 3-1 Graph and schematic illustrating seasonal differences in the quality of water from a public-supply well in California. Review of historic water quality data helped draw attention to the way in which the wellbore itself functioned as a preferential flow pathway, resulting in lower-quality water following times of light or no pumping.

Specific Capacity

The apparent specific capacity or ratio of the yield of each well to its drawdown, expressed in gallons per minute per foot of drawdown (gpm/ft), is used to plot the operational trend of each well. The apparent specific capacity for a well is determined by dividing the pumping rate (gpm) by the observed drawdown (ft) in the well measured at any specific time. Specific capacity depends not only on the transmissivity of the aquifer but also on well construction factors, such as screen type, well diameter, degree of aquifer penetration, and degree of well development. In general, when a well is new or new screens have been installed, the specific capacity is expected to fluctuate with pumping rate, surface water conditions, temperature, aquifer water levels, and degree of screen efficiency. Specific capacity needs to be monitored regularly. When specific capacity declines (Figure 3-2), it means something has changed, and maintenance or rehabilitation of the well is needed.

Differential

The differential between the pumping water level inside the well and the water level in the aquifer outside the well can be measured in any nearby monitoring or observation well. This value can identify changes in drawdown that relate to well production and efficiency or regional water-level changes. The differences in water levels should be measured from surveyed elevations and plotted with time. The differentials are affected by aquifer plugging from particle movement or biological factors and subsidence. If the monitoring well level decreases with time, it may suggest the regional groundwater is lowering, meaning the water being pumped is not sustainable.

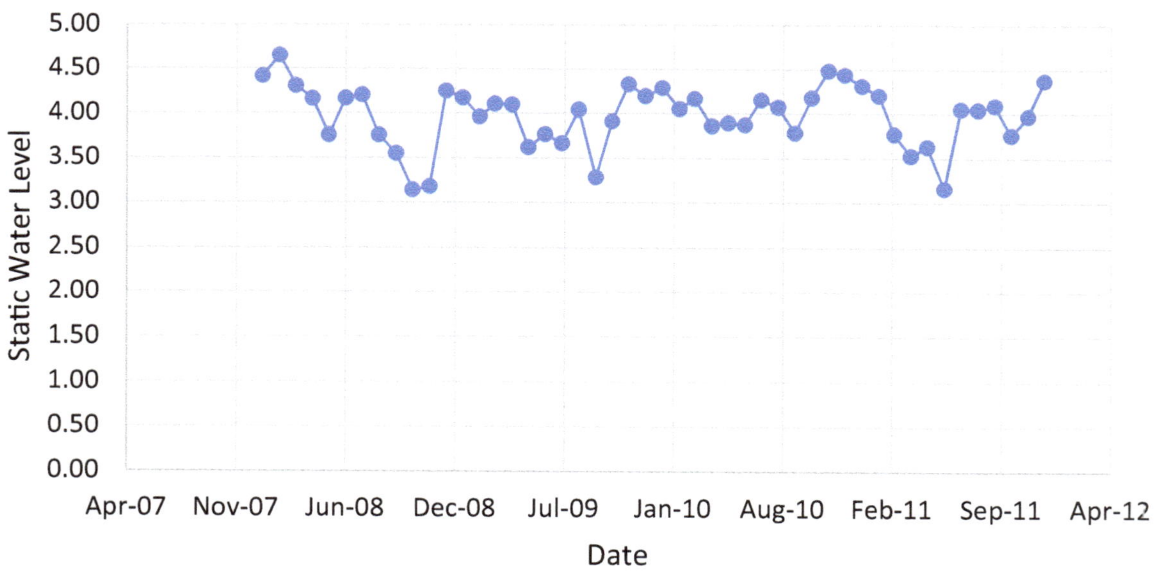

Source: Courtesy of Frederick Bloetscher

Figure 3-2 Specific capacity monitoring

WELL MAINTENANCE

Well maintenance activities should also be recorded. These records should include:

- the dates that maintenance was performed,
- results of pre- and post-maintenance pumping tests,
- methods (and materials) used in the maintenance procedures including well rehabilitation, and
- other factors such as the coloration of the pumped water, amounts of sand removed, odors, and water quality analyses.

These data can be used to predict times when maintenance needs to be performed, identify possible causes of well decline, and plan for annual budgets for wellfield management.

WELL ABANDONMENT

Records of any wells that are abandoned should also be maintained. These data should include:

- survey coordinates of the well,
- identification of markers or other indications of abandoned well location,
- the date the well was taken out of service,
- the date the well was properly abandoned,
- a description of the methods and materials used to abandon the well,
- reasons why the well was taken out of service, and
- historic performance records of the abandoned well.

WELL LOSSES

Drawdown values obtained for a single pumping well using the Thiem formulas represent the case where only the head losses suffered by water movement through the formation under laminar flow conditions are taken into account:

$$Q = \pi k \frac{h_2^2 - h_1^2}{In\left(\frac{r_2}{r_1}\right)} \qquad \text{(Eq 3-1)}$$

Where:

Q = volume

h = the distance from the bottom of the aquifer to the water level (two points; one can be the well itself)

K = hydraulic conductivity

r = the radius from the center of the well to the measuring points (one can be the well casing)

The actual pumping level of a particular well cannot be calculated without considering high velocities and turbulence losses during pumping. At and near the well face, fluid velocities may become large enough that turbulent flow conditions exist. The magnitude of turbulence losses varies with each well because of differences in formation characteristics, screen slot sizes required, degree of well development, well diameter, and quantity of water being pumped. There are so many unknown quantities involved in the calculation of these individual factors that they are usually lumped together under the heading of well losses.

One method of approximating the well losses for a particular well is to use the step-drawdown equation, as defined by:

$$s_w = BQ + CQ^2 \quad \text{(Eq 3-2)}$$

Where:

s_w = observed drawdown in the pumped well
B = the coefficient of formation losses
C = the coefficient of well losses
Q = the pumping rate

The values of B and C (formation and well losses) may be calculated if proper test data are available. To collect such data, the finished well must be pumped at three to five increasing rates for equal periods of time and the drawdown measured for each pumping rate. When a full-scale aquifer performance test is not conducted, however, a step-drawdown test can differentiate the observed losses in the pumping well. Additionally, this test makes it possible to quickly compare the magnitude of well losses to determine when a well needs cleaning or other repair work. Irregular increasing well loss with increasing pumping rates indicates unsatisfactory development of a new well or deteriorating aquifer or well conditions in an old well. Small regular increases in well loss or decreases in well-specific capacity due to the transition from laminar to turbulent flow in the aquifer are normal.

Examples of Operations Reports

For guidance purposes, the following pages include a series of blank reporting forms that might be useful to a utility: Figure 3-3; Figure 3-4 (a), (b), and (c); Figure 3-5 (a), (b), and (c); and Figure 3-6 (a) and (b). All can be filled out by hand or on a computer (in most cases using Excel®). Readers should feel free to use or modify such forms as may be useful to their system. Utilities should check with their regulatory agencies to determine specific form requirements.

Visual Tracking from Operations Reports

Although data need to be entered into the forms, visual understanding is much easier to interpret. As a result, all the water quality, water-level, and operations data should be graphed to observe long-term trends that might need further investigation or action. Figure 3-2 is a graph of specific capacity. The dips indicate a need to redevelop the well. Once the redevelopment has been done, the operations staff can observe that specific capacity returned. If the dips occur often enough, there may be other factors at play such as biofilms. However, there is a need to ensure the meters are read correctly. Figure 3-7 is an example of a meter reading certification report.

Figure 3-8 illustrates the long-term water level for the well. If the water level declines, it may indicate that the aquifer does not recharge at the rate of pumping and that other users are pumping from the same source or anomalous weather patterns (drought) have interfered with recharge. Understanding what is influencing groundwater decline is imperative to good long-term decision-making. Figure 3-9 shows a long-term change in arsenic concentration. The upward concentration is indicative of leaching of pollution contaminating the well. Treatment will be required, but the source should be sought out.

Figure 3-10 shows a graphic of a long-term water pumping report. Increasing pumpage indicates added demands but should be compared to the long-term water levels. If pumped flow rates increase while water levels decline, overpumping is indicated. That is an unsustainable source.

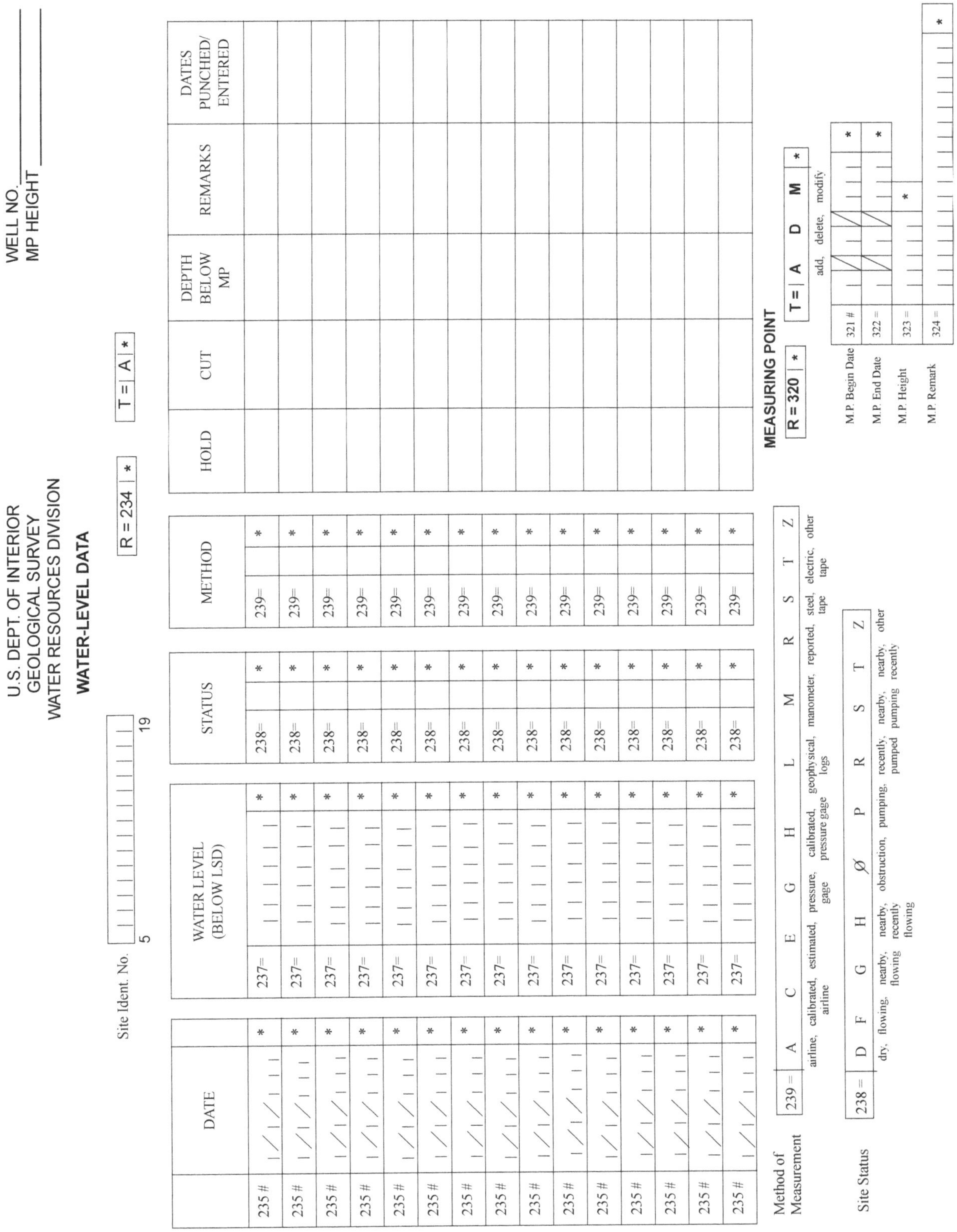

Source: USGS 1995

Figure 3-3 Water-level data report to monitor changes to long-term sustainability

WELL INVENTORY FORM **Send results of drilling/sampling to owner? Y N**

NAWQA Study Unit: ______________________ County: ______________________

Recorded By: ______________ Date: __________ Time: ________ Photo # _______ Roll # _____

Quad sheet: ______________ Contour interval _________(ft) Scale: __________ Year revised: ________

Sources of data: ____

WELL SITE INFORMATION

Site ID (C1): ____________________ Station name (C12): ______________________

Lat (C9): ____________ Long (C10): ____________ Land-surface altitude (C16): __________ft NGVD

Site accessible to vehicles? YES___ NO ___ Remarks ______________________________

Use of site (C23) ________; Use of water (C24): 1st________ 2nd________ 3rd________

OWNER INFORMATION

Well Owner Name (C161): ____________________ Phone: (H) ___________(W) ___________

Address: ____

____________________________ Zip: ______________

Tenant: ____________________________ Phone: ______________

Address: ____

____________________________ Zip: ______________

Previous owner: ____

Permission given by: ____

Permission to remeasure/sample/drill: YES____ NO____ CALL____ STOP BY____ OK IF NOT THERE____

Dates not available: ____

Owner: interested_____ neutral_____ not interested_____ Remarks ______________________

SITE CHARACTERISTICS (When well sampled, Land-Use Land-Cover form must also be filled out)

Land use: Urban___ Suburban___ Rural___ Crop___ Pasture___ Natural___ Other ______________

Potential contamination sources near well (septic systems, barnyard, feedlot, pasture, nearby fertilized fields, local storage of chemicals, other): ____________________ None visible ______________

Domestic wastes to: Septic tank___ Sewer___ Other ______________________________

Source: USGS 1995

Figure 3-4(a), (b), and (c) Well and site inventory form (*Continued*)

WELL INSTALLATION AND CONSTRUCTION

Method of well completion: Primary filter pack______; Secondary filter pack______

Type of annular seal__________________; Type of surface seal (C67);________________;

Protective casing____ (locked___, unlocked___)

Method of well installation (C65):_________________ Type of finish (C66):_______________________

Date well constructed (C21):_________________ Driller:______________________________________

Depth of well (C28):_________ft Depth to bottom of casing (C74):___________ft

Casing diameter (C79):_________ in. Casing material (C80):_________

If PVC used in construction, was glue used? YES_____ NO_____ Remarks ___________________________

Primary aquifer (C714) _____ Source of information: geology map, topo map, outcrop, drilling log, other _____

Method of well development (C69):____

Well-construction integrity checks: Date(s)_______________________ Type:_______________________

Pump Type:_____________________ HP:__________ Capacity:_____________________________

Pumping during normal use (if data available)

Approximate frequency (daily, monthly, yearly) ___________________
Approximate pumping rate (gpm): Range ____________________ Mean __________________
Approximate duration (min, hrs): Range _____________________ Mean __________________

Comments:____

____

WATER LEVEL

Hold_______________ Hold_______________ Hold_______________

Cut_______________ Cut_______________ Cut _______________

= _______________ = _______________ = _______________

+/-mp_____________ +/-mp_____________ +/-mp_____________

Water level, in ft below

LS (C30 or C237) _______________ _______________ _______________

Water-level status (C238): ____

Source: USGS 1995

Figure 3-4(a), (b), and (c) Well and site inventory form (*Continued*)

SAMPLING INFORMATION

1. Can water level be measured? YES___ NO___ Why not?________________________
2. Can well be sampled? YES___ NO___ Why not?________________________________
3. Sampling tap location: upstream from a holding tank? _____ pressure tank? _____
   If yes, give size of tank _________ gal
4. Any water treatment upstream from sampling point? YES ___ NO ___ If yes, describe: _______________
5. Other notable well features, if any (pitless adapter, frost pit, missing or leaky cap, well house, chemicals stored near well): ____
6. Water Quality? Taste:______ Odor:______ Color:______ Remarks:______________________________
7. Minimum rate at which pump in well can be operated: ________________gal/min

WATER-QUALITY FIELD MEASUREMENTS

Location of sampling point____

00010 Temperature _______°C 00095 Specific Electrical Conductance __________µS/cm

00400 pH _____________ 00300 Dissolved oxygen ___________mg/L 00076 Turbidity___________NTU

Other ________________________

AVAILABILITY OF ADDITIONAL INFORMATION (refer to Well Information Check List (see Installation) and include in inventory file):

1. Water-level records? YES___ NO___ Remarks:____
2. Pumping records? YES___ NO___ Remarks:____
3. Water-chemistry records? YES___ NO___ Remarks:_____________________________________
4. Borehole geophysical logs? YES___ NO___ Type:______________________________________
5. Surface geophysical surveys? YES___ NO___ Type:____________________________________
6. Aquifer tests? YES___ NO___ Type:____
7. Geologic materials samples? YES___ NO___ Type:_____________________________________
8. Land-use records for well vicinity (for example, pesticide and fertilizer application rates):

REMARKS, SITE SKETCH, AND WELL-HEAD SKETCH (showing locations of sampling point, holding tanks, and so forth) (Township____________ Range________ Section_______Quarter______):

Source: USGS 1995

Figure 3-4(a), (b), and (c) Well and site inventory form

RECORD OF WELL DRILLING, CONSTRUCTION, AND COMPLETION

SITE ID______________________ STATION NAME____________________________ OTHER ID____________________________

7.5' QUAD__________________________________ COUNTY ________________________________STATE __________________

OWNER ____ DRILLER ____

WELL DRILLING

START DRILLING: DATE _______/_______/_______ TIME _______:_______ EST

COMPLETE DRILLING: DATE _______/_______/_______ TIME _______:_______ EST

EQUIPMENT/MATERIALS DECONTAMINATION PROCEDURES:

DETERGENT WASH _________________; STEAM CLEANED _________________; OTHER ______________

DRILLING METHOD:

______ AUGER (TYPE: ____________________________); ____ ROTARY (TYPE:______________________)

______ PERCUSSION (TYPE:______________________________); ____ OTHER ______________________________

BOREHOLE DATA:

BOREHOLE DIAMETER: _______________ inches; TOTAL DEPTH OF BOREHOLE: ________________ feet;

APPROXIMATE DEPTH TO THE WATER TABLE: _______________ feet

MATERIAL DESCRIPTION				FROM	TO	THICKNESS
SAND, SILT, CLAY ETC	SORTING	COLOR	WET/DRY	feet	feet	feet

Source: USGS 1995

Figure 3-5(a), (b), and (c) Record of well drilling, construction, and completion (*Continued*)

WELL CONSTRUCTION

START WELL CONSTRUCTION: DATE _____/_____/_____ TIME _____:_____ EST

COMPLETE WELL CONSTRUCTION: DATE _____/_____/_____ TIME _____:_____ EST

CASING/SCREEN MATERIAL	CASING THICKNESS, SCREEN TYPE, SLOT SIZE, ETC.	DIAMETER	FROM	TO	TOTAL LENGTH
		inches	feet	feet	feet
CASING:	-				
SCREEN:					

WELL COMPLETION

START WELL COMPLETION: DATE _____/_____/_____ TIME _____:_____ EST

COMPLETE WELL COMPLETION: DATE _____/_____/_____ TIME _____:_____ EST

COMPLETION ELEMENT	COMPLETION MATERIALS	AMOUNT	FROM	TO	TOTAL LENGTH
		lbs; lbs/gal; etc.	feet	feet	feet
PRIMARY FILTER PACK	-				
SECONDARY FILTER PACK					
ANNULAR SEALS					
SURFACE SEAL					
WELL PROTECTOR					

COMMENTS____

____

Source: USGS 1995

Figure 3-5(a), (b), and (c) Record of well drilling, construction, and completion (*Continued*)

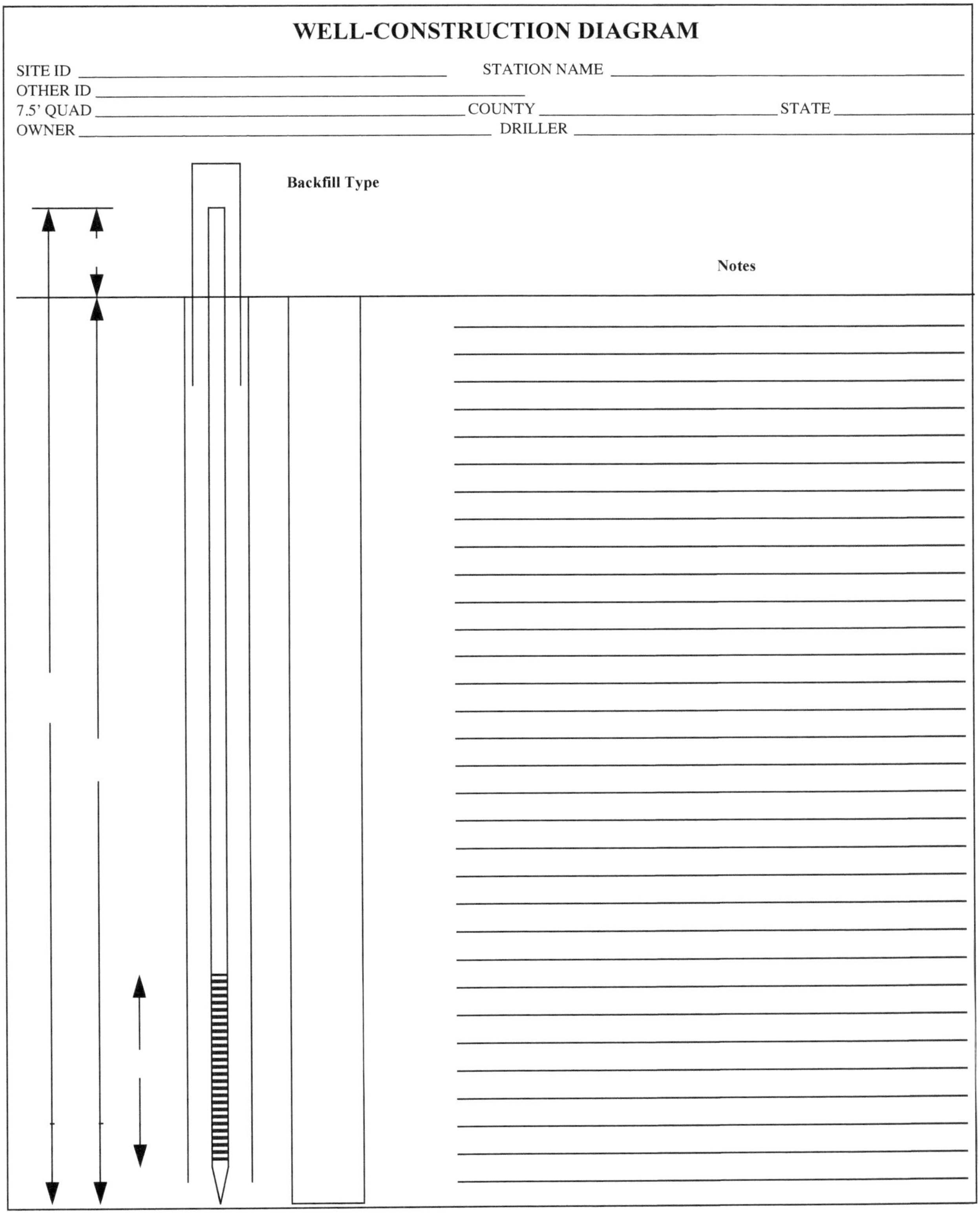

WELL-CONSTRUCTION DIAGRAM

SITE ID ____________ STATION NAME ____________
OTHER ID ____________
7.5' QUAD ____________ COUNTY ____________ STATE ____________
OWNER ____________ DRILLER ____________

Backfill Type

Notes

Source: USGS 1995

Figure 3-5(a), (b), and (c) Record of well drilling, construction, and completion

RECORD OF WELL DEVELOPMENT

Date: ____________________ By: ____________________

SITE ID____________________ STATION NAME______________________________ OTHER ID ________________
7.5' QUAD________________________________COUNTY______________________________STATE__________
OWNER____ DRILLER ______________________________________

WELL CONSTRUCTION

TOTAL DEPTH OF WELL __________________________ WELL DIAMETER __________________________
DEPTH(S) TO SCREENED OR OPEN INTERVAL(S) ____

WATER LEVEL

MEASUREMENT POINT (MP) DESCRIPTION ____
MEASURING POINT ______________ feet ABOVE _______ BELOW _________LSD (Land surface datum)

DATE	TIME	PERSON-NEL	TYPE - post drilling, pre-development, post-development,....	HOLD	CUT	H_2O DEPTH BELOW MP	MP	H_2O DEPTH BELOW LSD
				feet	feet	feet	feet above LSD	feet

ESTIMATION OF PURGE VOLUME AND PURGE TIME FOR WATER-QUALITY SAMPLING

Well volume = V = 0.0408 HD² = ___ gallons

V= volume of water in the well, in gallons
D = inside diameter of well, in inches
H = height of water column, in feet

Well casing diameter (D)	Gallons/foot of casing	Well casing diameter (D)	Gallons/foot of casing
1.0	0.04	6.0	1.47
1.5	0.09	8.0	2.61
2.0	0.16	10.0	4.08
3.0	0.37	12.0	5.88
4.0	0.65	24.0	23.5
4.5	0.83	36.0	52.9
5.0	1.02		

Purge volume = (n)(V) = ______ gallons

n = number of well volumes to be removed during purging

Q = Estimated pumping rate = ___ gallons per minute

Approximate Purge Time = (Purge Volume)/Q = ______ minutes

Source: USGS 1995

Figure 3-6(a) and (b) Record of well development (*Continued*)

RECORD OF WELL DEVELOPMENT--Continued

Date: ____________________ By: ____________________________

SITE ID____________________ STATION NAME__________________________________ OTHER ID ________________

WELL DEVELOPMENT

METHOD(S): BAILING___; MECHANICAL SURGING ___; INDIRECT EDUCTION ___ ; PUMPING/OVERPUMPING, AND BACKWASHING___ ; BACKWASHING___ ; JETTING WITH WATER ___; JETTING WITH AIR___; OTHER____
PUMP DESCRIPTION____

TIME	TEMPER-ATURE	CONDUC TIVITY	pH	DISSOLVED OXYGEN	TURBIDITY	APPROX. PUMPING RATE	COMMENTS (INCLUDING CLARITY OF WATER AND SUCCESS OF DEVELOPMENT,)
HR:MIN	°C						

REMARKS:____

Source: USGS 1995

Figure 3-6(a) and (b) Record of well development

PC CONTROLS INC

P.C. Controls Inc
132 NE 30th Street, Wilton Manors, FL 33334
Phone: (954) 568-9663, Fax: (954) 563-0919
E-mail: pccontrolsinc@bellsouth.net

Certified Flow Meter Test Record

Customer: City of Dania Beach

Consumptive Use Permit # 06-00187-W

Test Site Influent to Membranes Flow transmitter, FIT-1929

District ID# ___________

Contact Nate Costa Number 954 924-3616

Meter Model Rosemount 8750WA Serial# 0011439

Meter Accessories: Mag

Pipe Information at test site

Material Stainless Steel Schedule 10

Outside Diameter 10.75" OD Pipe Thickness .165" WT

Transducer Spacing 7.78" V-Mount Distance from Meter installation is 5 feet

Test Information using Fuji Porta-Flow Portable Ultrasonic Flowmeter

Test performed @ ___777.5_____________ GPM

Initial Meter Reading at Calibration Start: _1955916032_X _1_ End: _1955924096___X_1____

____8064______Gallons Totalized in _631_____Seconds Average _766.8__ GPM

Test Meter Reading at Calibration Start: ____0_______X_1__ End: ___8008_______X_1____

____8008______Gallons Totalized in _618_____Seconds Average _777.5__ GPM

Meter Tested at _99.9_ % Accuracy Last Meter Test _6/20______ Last Accuracy_100.1%_

Comments: Prior to calibration meter was reading 99.9% of test meter – Site meter read 766.7 GPM and test meter read 777.5 GPM.

Performed By: R Felicien Date: 6/3/21

This is to certify that the above listed equipment has been calibrated and meets or exceeds all published specifications. Calibration was performed using standards and/or instrumentation whose accuracy is traceable to the N.I.S.T.

Signature :

Source: Courtesy of City of Dania Beach (Fla.)

Figure 3-7 Certified flow meter test record

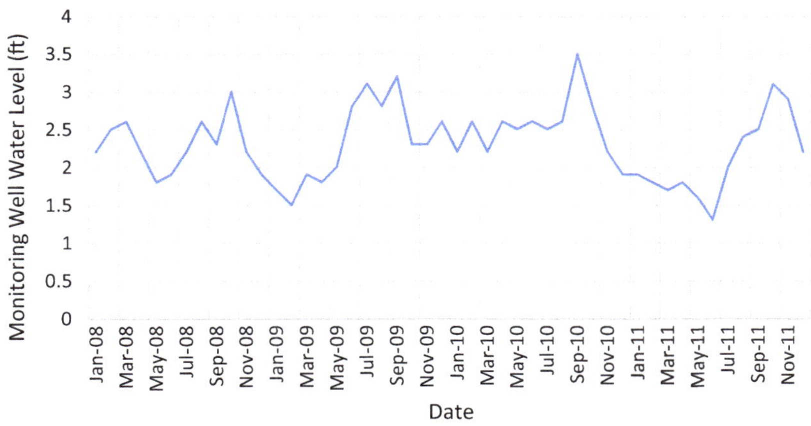

Source: Courtesy of Frederick Bloetscher

Figure 3-8 Long-term water-level tracking chart for a monitoring well

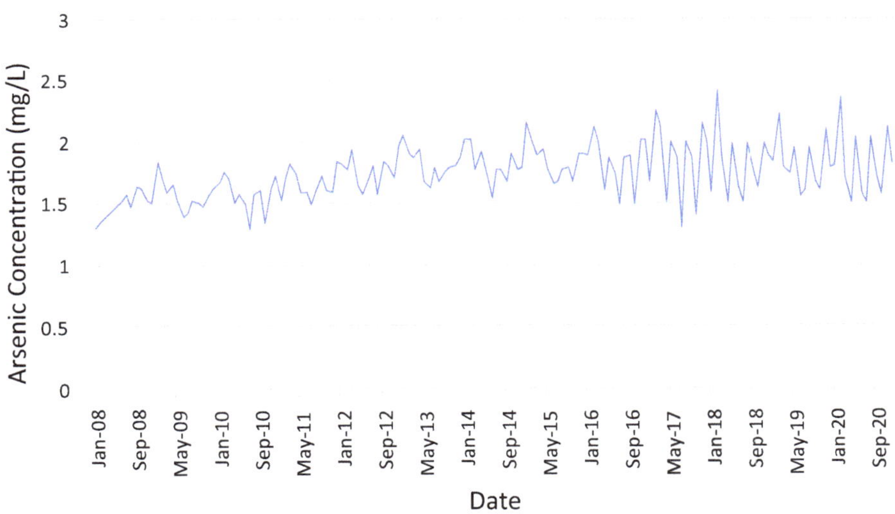

Source: Courtesy of Frederick Bloetscher

Figure 3-9 Long-term water quality tracking chart (arsenic)

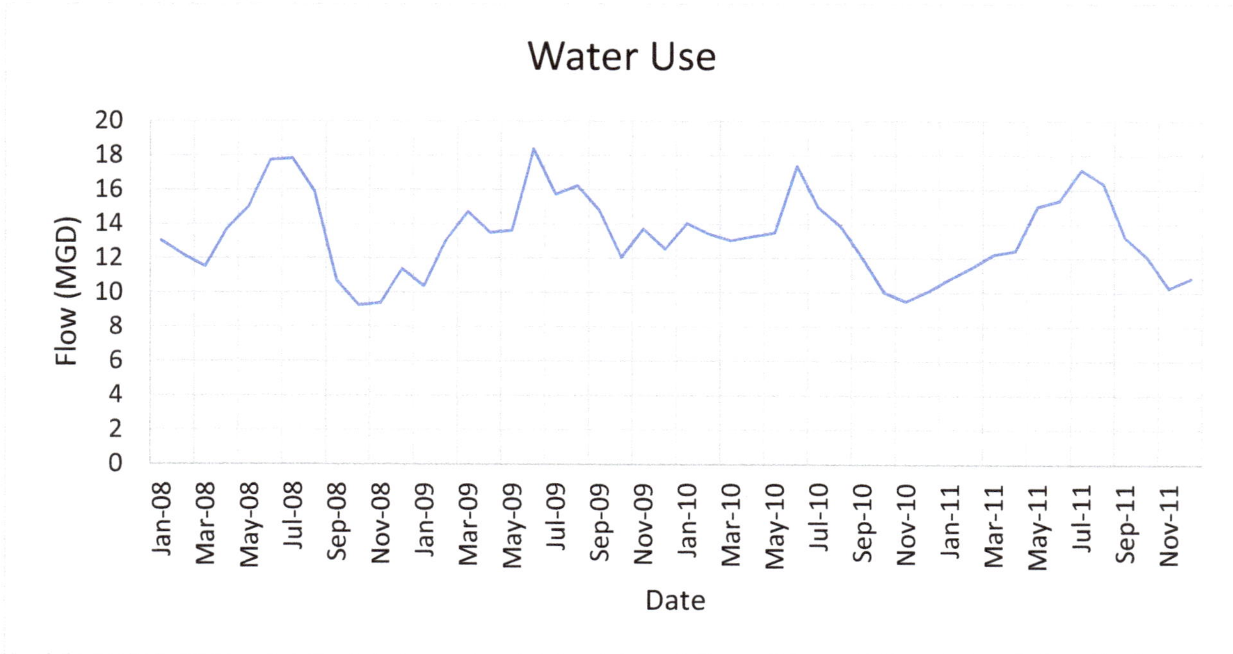

Source: Courtesy of Frederick Bloetscher

Figure 3-10 Long-term water pumping report

SUMMARY

If original records have been lost or not kept at all, manufacturers and well drillers (who maintain itemized records, including details for original pumps sold and installed) can be contacted. These records should be collected and filed or incorporated into a binder system. Copies of those records give indications of the original pumping levels and head conditions. They may also indicate general area changes in water levels. Actual drilling logs may be replaced by running gamma-ray logs in both new and old wells and then compared to correlate formation compositions. Changes in water quality may be documented in state health agency files and be relevant to well maintenance or used to identify a previously unknown vulnerability to contamination. "Retroactive" historic data of local aquifer conditions may also be available through records maintained by regulatory agencies.

REFERENCES

Eberts, S.M., M.A. Thomas, and M.L. Jagucki. 2013. "The Quality of Our Nation's Waters—Factors Affecting Public Supply-Well Vulnerability to Contamination—Understanding Observed Water Quality and Anticipating Future Water Quality." US Geological Survey (USGS) Circular 1385. http://pubs.usgs.gov/circ/1385/ (accessed 2/2/22).

Jurgens, B.C., P.B. McMahon, F.H. Chapelle, and S.M. Eberts. 2009. "An Excel Workbook for Identifying Redox Processes in Ground Water." US Geological Survey (USGS) Open-File Report 2009–1004. Reston, Va.: USGS. http://pubs.usgs.gov/of/2009/1004/ (accessed 2/2/22).

USGS. Variously dated. *National Field Manual for the Collection of Water-Quality Data*. Book 9 of *USGS Techniques of Water-Resources Investigations*. Reston, Va.: USGS. http://pubs.water.usgs.gov/twri9A (accessed 2/2/2022).

USGS. 1995. Ground-Water Data-Collection Protocols and Procedures for the National Water-Quality Assessment Program: Selection, Installation, and Documentation of Wells, and Collection of Related Data, USGS Open-File Report 95-398. Reston, Va.: USGS.

This page intentionally blank.

Chapter 4

Operations Problems, Well Plugging, and Methods of Correction

Wells usually provide relatively consistent, maintenance-free water supplies over many years. Most operators are familiar with mechanical failures of pumps and motors caused by wear, age, or lightning, but in many cases, problems can be a symptom of fouling or plugging. These problems can create excessive drawdowns, loss of specific capacity, and other operational inefficiencies.

All wells can plug or foul as a result of hydrogeologic, biologic, geologic, engineering, and construction factors. The source of problems is usually physical, mechanical, or environmental. This chapter provides a basic discussion of well performance and fouling problems, their causes, options for correction, and the economics of rehabilitating old wells. Three case studies are included at the end of the chapter that discuss fouling problems in detail, and corrective actions that were, or were not, taken.

EVALUATING WELL PERFORMANCE

Specific capacity may be the most informative single factor in well performance (recall Figure 3-2). The specific capacity is the pumping rate per foot of drawdown. When plugging problems occur, the drawdown increases; therefore, the specific capacity decreases, although the total yield of the well may not decrease significantly.

A related factor is the entrance velocity at which water passes through the well screen or the edge of the formation (depending on the type of well). As the entrance velocity increases, sand, silt, and colloidal matter can enter the flow stream. Generally, entrance velocities are designed to be less than 0.1 ft/s (0.3 m/s) on the assumption that flow is laminar to minimize:

- turbulence around the well screen,
- precipitation of iron, manganese, and calcium, and
- particulates at the well screen.

In addition, a low entrance velocity prevents pressure changes in the well and releases dissolved gases. Conversely, higher entrance velocities may reduce some fouling problems.

A third set of indicators of well performance change are well and aquifer (nonlinear and linear) loss components of drawdown, which are derived during analysis of step-drawdown pumping tests (Krusemann and de Ridder 1994). Changes in these loss components can be used to determine whether well clogging is deep-seated (formational) or surficial (gravel pack and screen).

POOR WELL PERFORMANCE

Several performance problems are caused by fouling or sand and silt production in wells. These problems and their likely causes are as follows (Borch et al. 1993):

1. Water-level decline in the well
   a. Reduced hydraulic efficiency in the well, most commonly plugging or incrustation of the borehole, screen, or gravel pack
   b. Regional water-level declines
   c. Well interference or plugging of a gravel pack by sand, silt, or clay
2. Lower specific capacity
   a. Drop in pumping water level
   b. Reduction in pumping yield due to microbiological fouling, chemical precipitation, formation plugging, pump corrosion, and biofouling
3. Lower yield
   a. Dewatering or caving in of a major fracture or other water-bearing zone
   b. Insufficient development
   c. Lack of connection to water-bearing fractures
   d. Pump wear
   e. Impeller detachment from shaft
   f. Microbiological fouling, plugging, or corrosion and perforation of column pipe
4. Sand/silt pumping
   a. Presence of sand or silt in fractures intercepted by well-completed open hole
   b. Leakage around casing bottom
   c. Inadequate screen and filter-pack selection or installation
   d. Screen corrosion
   e. Collapse of filter pack due to excessive vertical velocity and washout
   f. Insufficient development
5. Silt/clay infiltration
   a. Inadequate seal around the well casing or casing bottom

b. Infiltration through filter pack
   i. "Mud seams" in rock
   ii. Insufficient development

Many of these performance problems can be traced to inadequate design and/or construction of the well. Several operational conditions that are warnings of design problems are overpumping (which results in lowering of the water table), clogging or collapse of a screen or perforation of a screen section, corrosion, incrustation, and wear aggravated by excessive intake velocities. Other design and construction errors include:

- poor selection of materials that lead to significant corrosion or collapse;
- incorrect specification of pumps;
- poor construction (casing cracks or leaks, leaking or missing grout, and misplacement of screens and gravel pack and misalignment); and
- lack of well development (poor well yield, turbidity and sand pumping, biofouling, incrustation, and excessive drawdown).

The well production rate is usually determined by the hydrogeologist on the basis of aquifer tests performed at the time of drilling (or shortly thereafter). The hydrogeologist's recommendation should be respected. Overpumping an aquifer can damage the well by reducing the storage and production capacity of a groundwater system. Wells that are too close together increase drawdown and pressure loss in the formation. In compactible granular formations, the water-bearing formation will consolidate. As a result, compaction and consolidation lead to a lower water table, less water storage space, reduced yield from individual wells, and potential collapse the well casing. Subsidence can follow.

COMMON PUMP OPERATING PROBLEMS

Breaking Suction

No pump should operate at a rate at which it breaks suction. Besides damage to the pump, the water level fluctuates violently when a pump breaks suction. The fluctuation creates a surge in the well and in the water-bearing formation outside the well. The force may collapse the well if it is not properly stabilized. A surge can also cause sand pumping. With the loss of suction, air is entrained with the water and causes it to appear milky. Air bubbles may also damage the distribution system piping by causing air pockets. A surge can dislodge corrosion products, slime layers, or other incrusted materials from the inside of the column or transmission pipes.

If a pump does break suction and pumping stops, the discharge must be reduced by partially closing the discharge gate valve until the pumping level in the well remains above the pump bowls. Closing the valve increases the head loss in the system, causing the pump to work against a greater total dynamic head and decreasing flow. This procedure also wastes power. To regain efficiency, the pump is usually set deeper (if conditions permit) or one bowl and impeller is removed from the pump assembly to change the operating characteristics of the pump.

Causes

A lower pumping level in a well that has previously operated satisfactorily may result from the following:

- The water table (nonpumping level) near the well may have dropped so that the pumping level was correspondingly lowered.

- The intake portion of the well may have become clogged with encrusting material, so greater drawdown had to be created to cause water to flow from the formation into the well at a given rate.

Lowered water table. The aquifer level near a well may recede seasonally or during long dry periods when recharge to the aquifer is at its minimum. An aquifer level is reduced if pumping is gradually depleting the stored groundwater. The successive installation of additional wells in an area with overlapping cones of depression can also cause the water table to recede. A receding water table will cause significant mutual interference between wells, and the overlapping cones of depression would reduce the water levels of the wells. Consequently, water levels in the aquifer will be lowered to a point that is lower than that found in a single operating well. Well pump curves are needed to understand the challenge.

Figure 4-1 illustrates the operating problem that results from a drop in the water table caused by any of these occurrences. Curve 1 represents the relationship between well yield and pumping level. Curves 2 and 3 represent lower pumping levels caused by recessions of the water table. The drawdown in each case is the difference between the depth to water at zero discharge and any other point on one of these curves. The limiting yield occurs where the increase in yield ceases to be approximately proportional to the increase in drawdown. As points A and C show, the lower water level reduces pumpage.

With a pump operating at point A on curve 1 and a drop in water level that changes the well performance to curve 2, the same pump may operate at about point C. This undesirable situation can be improved only by cutting back the pumping rate to less than 800 gpm (3,000 L/min).

Clogged intake. When a pump breaks suction, the water level fluctuates violently, producing a surge in the well and in the water-bearing formation outside the well. In very soft sandstone or a screened well not completely stabilized during construction, the surging action may dislodge fine materials and cause sand pumping.

If suction is lost, air becomes entrained with the water, causing it to appear milky. The discharge pulses, plus expansion of the entrained air bubbles, can reach the distribution system. Corrosion products or other incrustations can be dislodged from the inside of pipes and color the water. A loss of suction will interrupt the cooling effect of the water flow, resulting in the pump bearings heating up and eventually seizing.

Solutions

To determine whether the groundwater levels are decreasing as opposed to well-specific issues, monitoring wells should be installed and the data tracked with time. Within a relatively short period, the trend in groundwater levels should be obvious. If the groundwater level is declining, the withdrawal rate needs to be reduced and the assumptions used by the hydrogeologist to determine the pumping rate should be reviewed and adjusted. Lowering groundwater levels means that the withdrawals exceed recharge and the pumping rate is not sustainable.

For well-specific issues, water levels in all wells should be measured and recorded periodically. A continuing record of both nonpumping and pumping levels should be maintained. If the pumping level recedes in any well, the cause should be determined. Increasing drawdowns are an indicator of well issues. Clogged screens are one issue that will cause a suction break.

The pumping equipment should be adjusted if danger of breaking suction becomes apparent.

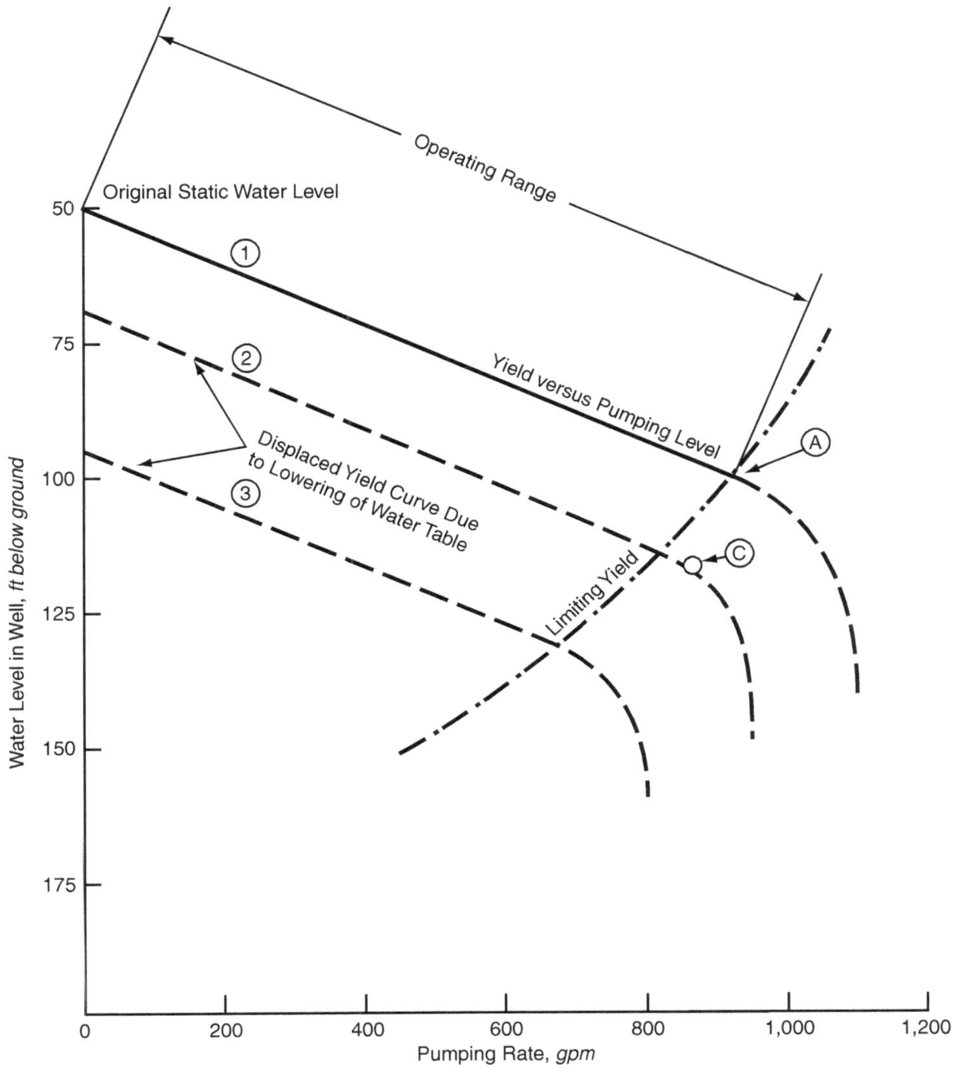

Figure 4-1 Operating problems resulting from a drop in the water table

PHYSICAL CAUSES OF WELL DETERIORATION

Water supply operators should evaluate any well failure or long-term decline in performance to determine whether physical or mechanical problems are causing the decline. A review includes:

- long-term drawdown trends,
- changes in viscosity or water temperature from the baseline data,
- pumping that exceeds the recharge rate,
- interference of other wellfields that are lowering the water levels,
- pumping that exceeds the design capacity, and
- nonuniform flow through the well screen.

The only way to determine whether any of these factors is occurring is to monitor the aquifer levels and to perform specific capacity analyses on a regular basis. The specific capacity should be determined at least annually on each well but analyzed only after the well is allowed to fully recover. Then, the well should be run for an hour to determine specific capacity. The result is then compared to the original data and plotted to show trends.

Evidence of a physical problem when the specific capacity declines includes:

- increased air in the discharge,
- increased sand, colloids, or other turbidity in the water pumped,
- excessive wear on the pump, and
- increased color or settlement around the well, which may indicate large quantities of sand removal.

Particulates, incrustation, calcium carbonate, and corrosion are the main causes of these visible changes.

Particulate Plugging

Sand, silt, and colloidal matter drawn into a well or well screen will plug the screens, providing less area to draw the water. Plugged screens increase the entrance velocity of the raw water, which can increase particle movement as well as drawdown. Particulates in the water tend to wear out well screens, increase pump wear, and decrease water quality.

In most cases, particulate plugging is caused by poor well design, including insufficient development of the well or inadequate formation sampling that leads to poor screen location. In some cases, the logging may not have been sensitive to thin layers of sand, silt, and colloidal matter that now may be exposed. In wells with gravel packs, incomplete development or overpumping may be indicated by plugging of the gravel pack and the screens. Consistent production of sand or silt in a well can collapse the formation above the screen or, worse, at the surface. Particulates also form a nucleus for chemical incrustation on a screen or column.

Plugging by Iron and Manganese

At a pH less than 5.0, iron and manganese remain dissolved as Fe^{+2} and Mn^{+2} in the water supply. Depending on the redox potential, they may remain in solution at higher pH, which is why most membrane systems add acid to keep precipitation from occurring (in addition to preventing calcium carbonate deposition). In the dissolved form, these metals can collect around the well screen in an insoluble mass. The presence of 2–3 mg/L dissolved oxygen (O_2) or a higher pH is a greater concern because ferric or manganic compounds can precipitate on the well screens. This precipitation will form nodules that collect additional ferric or manganic precipitates. Oxygenated water (used to lubricate pumps) or cavitation (caused by turbulent flow) aggravates iron precipitation and encourages the growth of iron-related bacteria around the well screen.

Acidic groundwater (pH less than 7.0) may dissolve calcium carbonate from the formation materials, causing them to move into the well screen or increase turbidity. The reduced pressure in the well and gravel pack due to pumping causes precipitation and scaling of calcium carbonate and may encourage iron and manganese precipitation.

Corrosion

Three general types of corrosion involved in water wells are hydraulic, chemical, and galvanic. Hydraulic corrosion is caused by turbulent flow, hard particulates, or wearing flow velocities, which abrade well components. Hydraulic corrosion is generally because of particulate matter from incomplete well development or fine material within the formation that is not screened out. Corrosion enlarges screen slots or open holes in the casing, which allows even larger particles into the casing, and the deterioration accelerates. In time, the casing material reduces and potentially collapses. Cavitation caused by turbulent flow aggravates corrosion by flaking off pieces of metal. Pumping at rates higher than design flow is the primary cause of hydraulic corrosion.

Chemical corrosion is the dissolution of metal, typically zinc or iron, into solution through carbonation or oxidation-reduction reactions. Chemical corrosion is more of a problem in older wells because of materials used in the past. Chloride ions that exist in raw water can form weak acids that attack metals. Sulfide ions also create acids in certain environments that may attack metal surfaces. Oxidation and reduction reactions occur in groundwater environments and can accelerate corrosion in a well. The presence of highly dissolved oxygen may accelerate desiccation of brass or other pipe.

Galvanic corrosion is caused by the generation of electric currents in dissimilar metals. Galvanic corrosion is often a problem with stainless steel pumps that are connected to steel column pipes with bronze centralizers in a steel casing. Newer technologies and the use of fiberglass, bronze, and plastics have reduced galvanic corrosion. However, the higher the conductance between two metals, the greater the potential for galvanic action. This corrosion is typically found where the casing screen is joined, where the submersible pumps are joined to the column pipes, or where bronze spiders exist. Poor pump alignment, stressed threading, or poor welds may encourage this type of corrosion.

Of increasing concern is the feedback of electrical currents from high-voltage power lines that create induced voltage on an underground pipe distribution system via the ground for water wells. This activity may be intermittent in effect. Because of soil resistance conditions, the effect may only be apparent during drought conditions, and progressive damage may develop over several years. It may be necessary to electrically isolate the pump from the distribution system and install a sacrificial anode or automatic electrical compensating device on the system.

Reducing Physical Causes

Proper design will reduce potential excessive entrance velocities or improper screen placement that can allow fine-grained formations to migrate into the wells. Proper materials such as plastics or fiberglass instead of steel or stainless steel can be used in many situations. Dissimilar metals should not be used in close proximity. Improper construction, poor grouting, excessive screen or casing damage, or the removal of protective sealants can lead to physical deterioration of the well. The improper application of certain chemical reagents, especially chlorine, and sequestering reagents or those used during redevelopment may exacerbate deterioration. Finally, overly aggressive pumping for redevelopment, overpumping of the system, or the improper use of surging may cause structural damage to the well in the long term. In all cases, the solution is to ensure proper construction to start and then monitor operations with time.

Biofouling/Microbiological Fouling

Aquifers are ideal environments for bacterial growth because there are nutrients, stable temperatures, and surfaces to colonize. The flow of water provides a consistent nutrient supply to the bacteria, and, except for the immediate pumping zone, the formation is not disturbed. The cycling of a pump on and off with the exchange of atmospheric air in the casing provides the oxygen supply to maintain low-aerobic conditions sufficient to maintain growth of certain bacteria in the well. All spaces within the formation are potential areas for colonization. Vugular formations (rock with air pockets) or those with significant fracture surfaces are ideal for creating large biofilms within the aquifer but never indicate severe plugging because of the size of the organisms in comparison with the opening size. Hence, all aquifers should be considered as sources for bacteria, albeit not necessarily coliforms.

The presence of bacteria can lead to microbiological fouling of the well or the creation of a biofilm. A biofilm is an entire active ecosystem, providing an environment for survival to a variety of microorganisms by storing and transporting nutrients. The bacteria absorb nutrients such as iron, nitrogen, and oxygen within the biofilm walls—leading to the formation of tubercles and films that reduce the capacity of the pumps and casings while protecting the bacteria cells from external reagents such as chlorine but trapping iron, sulfur, manganese, and other nutrients. Wells with water-lubricated pumps are subject to severe microbial action.

Microbiological fouling is generally interrelated with physical and chemical processes. Microorganisms encourage changes in the electrical potential and pattern of the well surface by creating anodes on metallic surfaces to transfer ions that can incrust or corrode the well and piping materials, enhancing physical and chemical deterioration. The most commonly described bacteria are "iron bacteria," which tend to precipitate iron and create a characteristically rust-colored slime and hard tubercles. The precipitated iron creates the hard iron oxide tubercles by removing iron from the casing surface in other areas. It can be assumed that all well deterioration problems include some degree of microbiological fouling.

Iron bacteria are not the only challenge. Sulfur-reducing bacteria (SRB) reduce sulfates or elemental sulfur to sulfides by respiration. The mineralization of iron and sulfur (forming Fe II sulfides) caused by SRB with low redox conditions will cause a slime matrix that is generally black in color. This process also commonly produces hydrogen sulfide. SRBs also interact symbiotically with the aerobic iron bacteria. Meanwhile, sulfur-oxygenating bacteria form light-colored filamentous slime biofilms where sulfides are present in groundwater. An example is the *Thiothrix* genus of bacteria. Precipitates of the iron, sulfur, and manganese occur within the biofilm, so biofilms can eventually clog screens.

Another challenge is the slime-forming bacteria, which come in many varieties. They can live symbiotically with other bacteria by providing the protective slime coating that they secrete. Some of the more common species of the slime-formers are those in the *Pseudomonas* genus, some of which are opportunistic pathogens. As biofilms grow, they may form filaments that serve to reinforce the biofilm and protect the underlying bacteria.

The typical symptoms of microbiological fouling problems are:

- a decrease in the water quality,
- increased drawdowns,
- reduced specific capacity,
- a change in the amount of iron or manganese in the water supply, and
- an apparent increase in microbiological densities such as an observance of slimes or staining from the raw water.

Several steps should be followed to look for bacteria. A downhole camera should be used to look for the visible signs of biofilm deposits. Any equipment that is pulled out of the wells should be thoroughly cleaned so that other wells are not contaminated.

Biofouling can be detected early via indicators of increased bioaccumulation. Since the early 1990s, relatively effective and simple-to-use field tests of biological activity have been developed. Among those available are relatively simple-to-use cultural enrichment methods, including the widely used BART* and a similar method developed independently by practitioners in Argentina.† Their theory and use are summarized in Cullimore (2000, 1993) and Gariboglio and Smith (1993), respectively. Evaluations of BART methods and comparison with other culturing methods by independent users are available (SWWI 2000, Smith 1992). These tests are species-specific and require the collection of a water sample

* Droycon Bioconcepts Inc. (Regina, Saskatchewan)

† MAG Laboratorio Microbiologia Industrial, La Plata, Argentina

and exposure to the media used in the test. If present in the sample, the organisms will grow.

Culturing methods such as BART or MAG tests are limited to use in detecting bacteria that can be cultured (a minority of those present) and do not permit a direct observation of biofilm properties. Microscopy (which has its own limitations) provides the capacity for direct observation of biofilms. This is best conducted using relatively intact biofilm examples. For this reason, biofilm collection methods have been developed to aid in providing high-quality biofilm samples. Borch et al. (1993) and Smith (1992) describe many of these.

Collectively, such tests (plus some others) permit a qualitative evaluation of the microbial ecology of wells and water systems that is useful in planning rehabilitation and maintenance. Experience in their use for this purpose is evolving rapidly.

If BART shows bacteria are present, a microbiological laboratory capable of identifying bacterial and fungal species should be consulted. However, because most aquifer formations have thousands of microbes, the BART step could be skipped and water samples delivered directly to the microbiological laboratory for analysis. Repeat samples are beneficial because they indicate the changes that may be occurring in the formation. A common but unrecognized benefit of the test for heterotrophic plate count tests is that it can reveal additional bacterial strains—they are the pink dots that grow on the filter test media that are not coliforms.

When a biofouling problem has begun, little can be done to remove it permanently. Control of the colonies is the best mitigation strategy; the sooner biofouling is detected, the sooner treatment can begin. Methodologies to detect and manage biofouling through maintenance have improved significantly in recent years and are described in a variety of available literature. Chlorine, carbon dioxide (CO_2), and acids have been used to help control bacteria in wells.

Biological Contaminants

One concern with biological contaminants is the source, especially regarding pathogenic organisms. There are more than 100 microorganisms that are considered human pathogens (Feachem et al. 1981), most of which are introduced into the body via ingestion, inhalation, dermal contact, or entry through wounds or body orifices (Hurst 1996). Infected people excrete large numbers of these pathogens, which often find their way into ground and surface waste systems. Each organism has a different dose-response relationship with a vastly different threshold dose for infection. Typically, bacterial infections require very high quantities of these organisms, whereas with certain viruses, one organism is sufficient to cause infection. Fortunately, available studies indicate that bacteria are generally removed during wastewater treatment and disinfection, but depending on the treatment process used, viruses may experience only a 50 percent removal rate (Yates et al. 1987). Chlorine is designed to remove bacteria from drinking water.

Relevant classes of microorganisms. Microorganisms associated with waterborne disease can be broken into three groups: protozoans, bacteria, and viruses. Each has a unique environmental fate and effect characteristics in groundwater systems. The following sections summarize each genre.

Protozoans. Protozoans and their cysts are common in surface waters and are much larger than either viruses or bacteria. The cyst stage is an encapsulation that protects protozoans from harsh environmental conditions, such as drought in the case of duck botulism. *Cryptosporidium* and *Giardia lamblia* are the two protozoans most studied because of their presence in drinking water (generally unfiltered surface water) and their recent link to waterborne-illness outbreaks (e.g., in Milwaukee in 1993). *G. lamblia* is believed to be the most common protozoan pathogen present in surface waters. Its population appears to remain constant throughout the year in surface water impoundments (Rose and

Carnahan 1992). Neither *Cryptosporidium* nor *G. lamblia* appears to be a common groundwater problem except in those aquifers under the influence of surface waters. Both organisms are generally believed to be too large to move significant distances in groundwater systems. If protozoans are found, it means there are large amounts of secondary porosity linking a surface water body to the well. Those wells must be treated as being "under the influence of surface water," which means surface water treatment rules like giardia removal are required to be met.

Bacteria. Bacteria are the most widely distributed lifeforms on earth (Chapelle 1993). Chapelle (1993) notes that bacteria are extremely important to consider in groundwater projects because they inhabit virtually every subsurface environment, producing methane (CH_4) gas and consuming organic-rich soils. As noted in the previous section, bacteria should be assumed to be present in every well.

The key bacteria responsible for waterborne diseases include *Legionella, Pseudomonas, Klebsiella, Escherichia coli, Shigella, Enterobacter, Salmonella,* and *Vibrio cholerae*. Pathogenic bacteria are approximately 0.4–14 µm long and 0.2–12 µm wide, which means they are much smaller than protozoans, thus making it easier for them to move in the subsurface. Most of the pathogens belong to a classification called gram-negative bacteria, which are found extensively in subsurface situations.

Bacteria have their own enzyme equipment, and most are motile, which allows them to move in the subsurface. Bacteria reproduce by splitting into daughter cells, each of which continues to split, forming additional bacteria and eventually a biomass. The respiration ability of bacteria permits them to survive in soils and aquifers. There are three respiration types: (1) those bacteria that use inorganic chemicals to serve as electron acceptors such as oxygen, ferric iron, and sulfates; (2) those that are aerobic (requiring oxygen); and (3) those that are facultative anaerobes (capable of fermentation or using oxygen as electron receptors) (Chapelle 1993). The respiration mechanism is important because it affects the ability of bacteria introduced to colonize wells and the aquifer; it also affects the growth rate of bacteria indigenous to the aquifer as a result of the constituents introduced.

One of the most common opportunistic bacterial pathogens is *Pseudomonas aeruginosa.* It is a common infection in hospitals (causing pneumonia). *P. aeruginosa* is an extraordinarily versatile organism that will live in nearly any environment as it requires no specific vitamins, growth factors, or amino acids; it is a facultative anaerobe. However, the most important concern of this pathogen is its ability to create a slime matrix that encapsulates other bacteria and protects them from otherwise harsh aquifer conditions.

Commonly found bacteria in the subsurface include *Gallionella* and *Desulfovibrio. Gallionella* is an obligate aerobe that obtains energy by oxidizing dissolved ferrous iron to form ferric oxyhydroxides—meaning it will be a problem in wells constructed with steel materials (Chapelle 1993). *Desulfovibrio* is an SRB that uses hydrogen or simple organic compounds as an energy source and sulfates as the terminal electron acceptor, which leads to hydrogen sulfide (H_2S) gas formation (Chapelle 1993). Other bacteria may also colonize the slime matrix (Bloetscher et al. 1998).

Viruses. Viruses are molecular entities that have little or no enzymatic equipment, no energy capability, and no mechanisms for synthesis. Typically, they are 20–300 nm in size. They cannot reproduce themselves, requiring a host cell to multiply. Pathogenic viruses tend to be smaller than other viruses and can only be seen with an electron microscope: most are 27–70 nm in size and symmetrical. Most viruses tend to be resistant to chlorine, but they may be inactivated to various degrees during water treatment processes or by chlorine, bromine, ozone (O_3), ultraviolet (UV) light, and/or formaldehyde (Block 1989). Viruses are conserved at -4°F or –20°C (Block 1989).

All viruses are composed of nucleic acid and either RNA or DNA (but not both), which allows them to replicate in other cells, including bacteria—when they are called bacteriophages (Chapelle 1993). Viruses are obligate parasites, always searching for the correct host cell that will allow the virus to multiply. Viruses cannot survive or infect without such a host organism or cell (Chapelle 1993).

Transmission of viruses occurs in one of the following ways (in order of prevalence):

- fecal to oral pathway,
- person to person, and
- respiration.

Human viruses found in natural waters are almost always associated with fecal material eliminated from the bodies of infected individuals (Sellwood and Dadswell 1991). Therefore, virus concentrations in wastewater are high. More than a million plaque-forming units of viruses are eliminated per gram of fecal matter from infected individuals. The number rises to 10 billion for those infected with *Rotavirus* species (Yates et al. 1985). Survival of these viruses demonstrates that rotaviruses are sufficiently hardy to survive wastewater treatment and disinfection processes—thus the regulatory concern regarding these organisms in injection programs.

Major viruses of concern are hepatitis A, coxsackie, echovirus, Norwalk, small round structured virus (SRSV), rotaviruses, and reoviruses (Block 1989). Although vaccines may be available for some viruses, the wild strains never disappear from the environment, indicating that continued vaccinations are important (Bouwer 1991). Note that COVID-19 is now found routinely in wastewater samples, as is poliovirus. A summary of those viruses of concern and the number of studies conducted, as reported by Gerba and Bitton (1984), is in Table 4-1.

TREATMENT OF CLOGGING AND FOULING PROBLEMS

Many older wells were installed with methods no longer in use today and do not meet today's standards. In these cases, some work can be done, but the problems probably cannot be fully corrected. For many older wells, acidification—typically using hydrochloric acid (HCl)—can improve performance.

Table 4-1 Viruses of concern summary

Virus	Number of Studies
Poliovirus	3
Echovirus	31
Coxsackie A	23
Coxsackie B	6
Coxsackie types 65–71	4
Hepatitis A	1
Norwalk	1
Calicivirus	1
Astrovirus	1
Enteric coronavirus	1
Reovirus	1
Rotavirus	2
Adenovirus	37
Total	**112**

Adapted from Gerba and Bitton 1984

Heating chemicals increases their activity, which is valuable in cool groundwater, and increases penetration of chemicals into the formation. Less-concentrated solutions can be used. Additionally, heat aids in shocking and disrupting biofilms and aiding in their reduction. Heat-amended chemical treatments with redevelopment are among the *best* documented of more recent treatments.

Another method, carbon dioxide injection, uses gaseous carbon dioxide and liquid carbon dioxide under 100 psi of pressure. If liquid carbon dioxide is used, the water freezes in the formation, causing cracking and loosening incrustation. The formation may also crack and loosen, which can free the fractured zones or crack the bedrock formations and potentially increase yield. After the carbon dioxide is injected, the well is surged and redeveloped. A chemical treatment and development step is frequently used in addition to the CO_2 treatment. This method is not effective on biofilms. As with hydrogen peroxide (H_2O_2), carbon dioxide treatment may promote further biofilm development. This is a form of fracking, so regulations should be closely monitored to verify this method can be used.

Sonar jetting involves deploying a sequence of small blasting caps suspended in the screen or exposed borehole and set off, sending shock waves and gas through the screen and into the formation. The goal is to blast incrustations off the well screen, formation, and casing. After sonar jetting, surging and full redevelopment of the well must occur to remove all the excess debris. Acidification improves the process to some extent. Potential problems with this process may be the inability to get permits to blast and the potential damage to the casing or the screens. An example case history of combined Sonar-Jet®‡ and chemical treatment use is found in Ground Water Science (2000).

Other methods that show some promise in certain cases are suction flow control devices and inner sleeve installations within the casing using entrained air to reduce fouling.

Sand Pumping

Amounts of sand as little as 0.3 ft^3/mil gal of water can cause many operational problems. In addition to causing excessive wear in pumps and valves, control orifices can become plugged, water meters stopped, and sprinkler heads clogged.

In a few instances, sand pumping cannot be eliminated, even if a well is properly designed and constructed. Reduced pumping rate by increasing head through valving may give some relief. Where water is discharged from the well into a large tank or reservoir, the sand may settle out and should not cause excessive problems, except for pump and valve wear.

If sand pumping cannot be eliminated, a centrifugal sand separator, as shown in Figure 4-2, may be used. Water enters the body of the device at a tangent immediately below the baffle. The small radius and high velocity create a large centrifugal acceleration, which throws the sand to the side of the device. The sand falls down the side and is collected in the centrifuge tube, and the sand-free water flows out through the hole in the center of the baffle.

The flow is maintained at a constant rate, independent of the inlet pressure, with a flow control valve rated at 0.5 gpm (2 L/min). This patented flow control valve uses a rubber orifice that contracts with increasing inlet pressure. The valve is designed for a pressure variation of 15–150 psi (100–1,000 kPa).

At suitable intervals, the volume of sand collected is recorded, together with the number of hours of operation. From these data, the average sand concentration may be computed because the flow through the tester is known. Any significant increase in sand production is noted immediately, and corrective action can be taken before appreciable

‡ Water Well Redevelopers & Surveys, 4800 Via Del Corral, Yorba Linda, CA 92887.

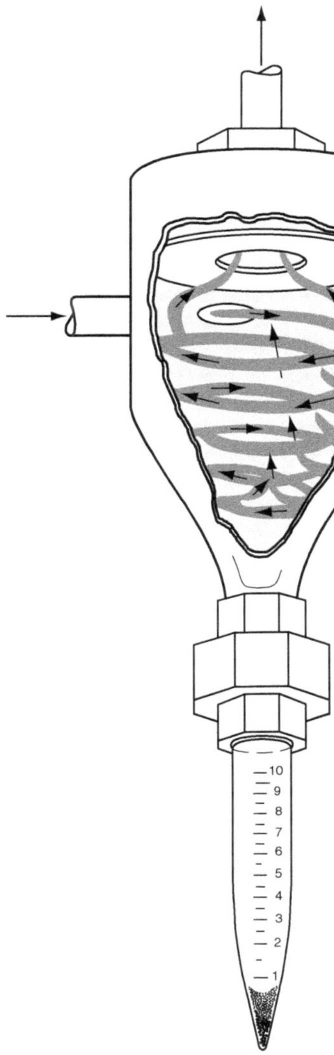

Figure 4-2 Centrifugal sand sampler

quantities of sand have entered the distribution system. One tester for each well suspected of producing excessive sand can be provided if economically feasible.

Sand flow into a well occurs only during turbulent flow unrestrained by the gravel pack or screen. Shifting the position of the pump suction, pipe, or intake or installing a suction flow control device between the screen and pump intake may be more cost-effective than a sand separator device.

If sand cannot be prevented, a sand separator may need to be installed (see Figure 4-3). Sand separators reduce sand going into the treatment processes or distribution system. If membranes are used, sand separators are highly recommended before cartridge filters to protect the integrity of the cartridge filters.

Biofouling

The goal in resolving biofilms is to disrupt the environment by chemical or physical means. Note that when a biofouling problem has begun, little can be done to remove it permanently. Control of the colonies is the best strategy; the sooner biofouling is detected,

Source: Courtesy of Frederick Bloetscher

Figure 4-3 Sand separator

the sooner treatment can begin. Methodologies to detect and manage biofouling through maintenance have improved significantly in recent years and are described in a variety of available literature.

Chemicals are commonly added to the well to enhance the effectiveness of redevelopment. Chemical choices for rehabilitation and maintenance cleaning depend on the clogging problem and the water quality and may differ in type and concentration, depending on whether rehabilitation or maintenance is the objective. For example, chlorine is used as a biocide for microbiological fouling, although in most cases it does not kill all the bacteria; it only controls the biofilm. A 12.5 percent sodium hypochlorite solution or calcium hypochlorite provides the chemical strength needed to control the bacteria. In some cases, hydrogen peroxide may be used to address biofouling problems, but certain bacteria (*Pseudomonas* species) may be able to use the oxygen to their benefit, increasing biological activity.

Another option is acidification, dropping the pH to less than 2. Hydrochloric (HCl), sulfamic (H_3NSO_3), acetic (CH_3COOH), glycolic ($C_2H_4O_3$), or nitric (HNO_3) acid is used, but these chemicals must be used carefully. Hydrochloric acid solution will remove or loosen incrustation in the screens or the column pipe, although it will not remove much biofouling. Deterioration of the well materials must be weighed against the removal of the biofilm or the incrustation. If biofouling removal is the objective, a solution of 10–15 percent acetic acid or 5–10 percent glycolic acid, amended with sulfamic acid to achieve a pH of 2, is commonly used. This solution attacks biofilm integrity without the aggressiveness of hydrochloric, nitric, or phosphoric acid solutions.

The addition of phosphates has been used, as it makes water "more slippery" and increases total well capability. However, the phosphates may *provide* a scarce nutrient for biofilm, and alternative chemicals that do not contain phosphates are now widely available. Guidance in chemical choice is supplied in Borch et al. (1993) and other publications, as well as on websites such as Ground Water Science. In all cases with chemical use, a plan for handling and disposing of hazardous material must be made. None of these chemicals should be discharged in an uncontrolled way to the ground; they must be hauled to an approved disposal site, such as a sanitary sewer.

Physical agitation or surging (redevelopment) is necessary to remove dislodged deposits during well treatments. At times, redevelopment may be used alone to remove incrustation or reduce fine material entering the well screen or gravel pack. Methods used are essentially identical to those used for original well development and are described in detail in Borch et al. (1993), Roscoe Moss (1990), Driscoll (1986), and other references on well construction and rehabilitation.

In cable-tool surging, tools are used to push water down into the well and pull it out, just as old hand-pump well systems worked. Initially, the surge device is operated at less than three strokes per minute at 6–10 in. (152– 254 mm) per stroke. Over time, the frequency and the stroke should be increased (up to about 5 ft/min [1.5 m/min]), which increases the surging. If the casing or the formation is weak or the screens are damaged, the well structure can collapse during surging.

Environmental Issues

Any of these methods creates potential environmental problems that must be addressed by appropriate regulatory agencies. All deplugging and cleaning methods require the discharge of water, which can contain chemicals, silt, sand, and other debris. The quality of this water may require treatment. A common method of disposal is to discharge the well volume into a tanker truck, stabilize the chemicals, and haul the debris to a wastewater treatment plant. Solids may need to be separated from liquids and disposed of separately in accordance with regulatory and wastewater plant requirements. The liquid may be hauled to a landfill in some jurisdictions. Appropriate regulatory agencies should be consulted and necessary permits obtained.

Ideally, minimizing the effect of treatment methods should be a part of the planning process. Actual chemical risk can be minimized by choice of chemical, using the least aggressive chemical concentration possible (relying more on development action and increasing effectiveness using heat), and handling and discharge risks are manageable with proper planning.

In each case, the waste stream characteristics must be identified, including:

- pH of the water,
- chloride level,
- toxic substances,
- silt,
- quantity of the water to be discharged,
- time element for which the discharge will occur (i.e., a relatively consistent flow over a period of time or surges),
- new water quality of the wells, and

- uptake of metals, synthetic organic chemicals (SOCs), or volatile organic compounds (VOCs) that might violate air or water standards.

Regulatory agencies that may be involved in any discharge to surface waters or wetlands include national, state or provincial, regional, and local agencies, especially for discharges to sanitary sewer systems.

ECONOMICS OF CLEANING PLUGGED WELLS

Cleaning a plugged well can help maximize the return on a capital investment in a well. Operation costs can be reduced by eliminating pumping inefficiencies and restoring the well efficiency to its earlier level. To properly evaluate the economics of cleaning plugged wells, historical records are needed for flow, test data, specific capacity calculations, development records, design details, pump performance curves, and some periodic information on inspections. These data will create a baseline to compare all future well performance calculations. The information may also be used in selecting proper methods for unplugging the wells. Routine monitoring should include flows, drawdowns, hours pumped, power usage, and calculated specific capacities, all of which should be plotted for each of the wells.

To evaluate the costs, the initial costs of service should be determined using the current pump and well performance. Changes in specific capacity will affect both the power and the hours pumped. The assessment over time indicates what the changes in the efficiency of the pump have been. Recordkeeping will tell operators and engineers when the well has deteriorated to a point that it needs to be rehabilitated. AWWA's Groundwater Resource Committee recommends tracking specific capacity over a 20-year period. A 15 percent loss in specific capacity over this period indicates efforts should be made to investigate and correct well issues.

The risks involved with well age must be compared with the cost to rehabilitate, the well's life expectancy, and the potential for gains in specific capacity to determine whether to rehabilitate the well. A well that has been rehabilitated several times and shows a general decline in performance and an inability to regain the initial specific capacity may be a candidate for abandonment (Figure 4-4), although a change in treatment method may reverse the trend. Any evaluation should include the cost to rehabilitate or to replace the well if the cleaning process is not successful, the value of additional water obtained, and comparison of the cost per unit of water pumped between treatment and other alternatives such as well replacement. Replacement wells have the added issue of the relocation of pumping, which may entail added property acquisition and, in some areas, water rights. Table 4-2 summarizes a fictitious example of how the replacement versus rehabilitation options must be weighed:

Table 4-2 Fictitious example of replacement versus rehabilitation

Well Rehabilitation	Replacement Wells
• Payback in water • Will break even in 1.8 years Cost: $20,000	• Exploration for a well • Permit application • Water rights acquisition • Design engineering • Land acquisition • Pipeline design • Treatment changes Cost: $300,000

Source: Courtesy of Frederick Bloetscher

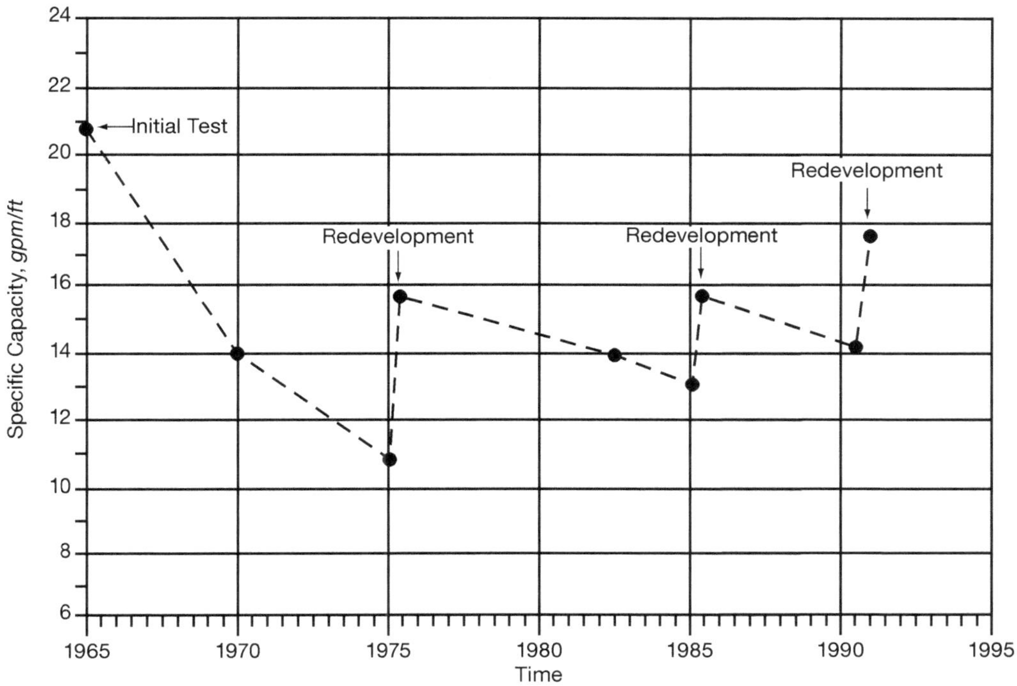

Figure 4-4 Well efficiency monitoring, Catamount Wellfield, well No. 54

When the current costs are developed and compared to the initial costs, and the risks are determined, a direct comparison of the current versus initial costs can be made. This comparison will indicate the change in operating costs due to inefficiencies of the well. A rule of thumb is that if there is a specific capacity decline of more than 15 percent, some rehabilitation should take place because the increase in operational costs will be high, although local conditions or costs may dictate other values. In most cases, there is a relatively short payback period between the costs of rehabilitating the well and the cost for installation of a new well. In cases when the wells are relatively old, the replacement cost should be used to determine payback period. For old wells where performance continues to decline, new technologies can provide substantial benefits. The following section is a series of calculations that have been made as an example.

Calculation

This analysis was originally presented at the AWWA Annual Conference in Toronto, Canada, on June 22, 1996, by Kenneth C. Gaynor with Hydro Group, Inc.

Example Well for Cost–Benefit Analysis

Original well performance

Static water level:	15 ft
Well depth:	85 ft
Available drawdown:	40 ft
Flow rate:	700 gpm
Pumping level:	48 ft
Drawdown:	33 ft
Specific capacity:	21 gpm/ft

Annual Operating Conditions

The volume of water pumped is 157 million gallons. Operational time is 3,738 h per year. The average kilowatt cost is 9 cents per kilowatt.

Calculation

Original Annual Operating Cost of Example Well

$$\$/\text{Hr} = \frac{\text{gpm} \times \text{tdh} \times 0.746 \times \$/\text{kWh}}{3{,}960 \times \text{pump eff.} \times \text{motor eff.}}$$

$$= \frac{700 \times 168 \times 0.746 \times \$0.09}{3{,}960 \times 0.82 \times 0.90} = \frac{7{,}895.6}{2{,}922.5} = \$2.70/\text{hr}$$

To get the cost per year, multiply by the hours:

= \$2.70/h × 3,738 h/year

= \$10,093/year

Example: Well Performance History

Well performance history is shown in Figure 4-5.

Current Situation

Specific capacity decreased to 7.5 gpm/ft
Maximum drawdown well: 40 ft
Maximum yield from wells: 300 gpm
Curve efficiency of 300 gpm: 60 percent

Operating Costs of Inefficient Well

$$\$/\text{Hr} = \frac{\text{gpm} \times \text{tdh} \times 0.746 \times \$/\text{kWh}}{3{,}960 \times \text{pump eff.} \times \text{motor eff.}}$$

$$= \frac{300 \times 198 \times 0.746 \times \$0.09}{3{,}960 \times 0.60 \times 0.90} = \frac{3{,}988}{2{,}138}$$

$$= \$1.86/\text{hr}$$

Current Costs

Given usage of 157 million gallons per year and to have 300 gpm, the pump must operate 8,722 h, which is \$16,223 per year.

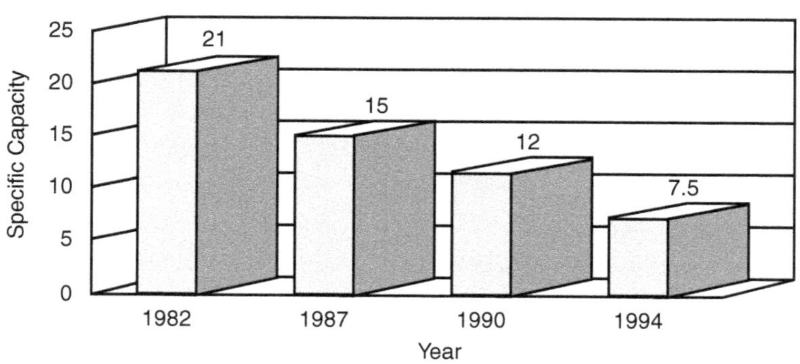

Figure 4-5 Example of well performance history

Potential Savings

Current well cost:	$16,223
Original cost:	– $10,093
Savings per year:	$ 6,130

Comparison to Cost of Redevelopment—Conclusions

The cost of redevelopment is $10,000; the payback is 1.63 years. The economic decision is to rehabilitate the well and repeat the redevelopment.

CASE STUDIES

In most well plugging situations, no single factor is involved with the well plugging or corrosion problem. As a result, no clear-cut methodology can identify and clear wellfield problems. Two of the following examples are from Florida; one system is a fresh, surficial aquifer zone, the other a deeper, brackish aquifer. Both have severe microbiological problems. The other case study is from the Midwest.

Collier County Wellfield

The Collier County Water Sewer District is in Southwest Florida surrounding the city of Naples. The service area is approximately 200 mi^2 (520 km^2). The system has grown since 1982 when the first wellfield was established.

The primary drinking-water supply is from the surficial Tamiami aquifer. Below the Tamiami aquifer is the Lower Tamiami aquifer, which yields higher-quality water for treatment. Whereas in some areas of Collier County there is little differentiation between these aquifers, at the district's wellfield, between 30 and 50 ft (9–15 m) of fairly tight clay and dolomite separate the two, limiting vertical recharge. The Lower Tamiami formation is highly transmissive and primarily made up of highly fractured and solutioned limestone. Recharge is primarily via rainfall, although a significant portion of the recharge may come laterally from the Big Cypress and Corkscrew swamp areas. The overlying area consists of low-density residential development (a minimum of 2.5 ac per household) and some minor incidental commercial development. Recharge capability is high. Below this formation is a series of progressively more saline aquifers, starting with the Hawthorn Group at 180 ft (55 m) below sea level.

In investigating potential well water supplies for the district, the Lower Tamiami aquifer was selected to provide the water, which is a highly transmissive production zone of vugular limestone. The production zone is between 60 ft and 140 ft (18–43 m) below sea level, and each well is designed for 1–1.5 MGD (3.8–5.7 MLD). Since the initial wellfield construction, two expansions have been completed. All of the wells are constructed with polyvinyl chloride (PVC) casings.

In 1990, the district designed a 12-MGD (45-MLD) membrane softening (nanofiltration) facility located 5 mi. north of its lime softening plant. The two plants manifold together and use the same wellfield for water supply.

Just before the design of the membrane softening plant, the district found several of its new 304 stainless steel column pipes had been sheared off only 18 months after installation. When fished out, the column pipes showed a black slime ringing the column pipe at the point of shearing. At the same time, a significant amount of slime was noted coming into the lime softening plant's degasifier towers along with significant amounts of hydrogen sulfide. The specific capacity of the wells was analyzed and found to be significantly reduced from their initial rates. The steel column pipes and other steel within the pumps and the facility showed a significant amount of corrosion. The typical corrosion was pitting, with black longitudinal slime running in the vertical direction of the column pipe (Figure 4-6). Analysis was taken of the slime and sent to Harco Technologies in Atlanta,

Figure 4-6 Example of corroded column pipe

Ga., and LayneAtlantic in Kansas City, Kans. Reports from these two companies indicated that there was a significant and widespread pitting of the column wall thicknesses, including some portions in which 75 percent of the column pipe's thickness had been lost. Bacteria consisting of *Gallionella* and other iron bacteria propagated on the stainless steel materials, and anaerobic SRB had developed a symbiotic relationship with the aerobic iron bacteria. Various slime-forming bacteria of the *Pseudomonas* genus had interjected themselves into the symbiotic relationship by providing an overlying slime layer to protect the iron and SRB. Because the buildup creates an anode on a pipe, it exacerbates the deterioration of the ferrous material. Unfortunately, these species are persistent, especially the pseudomonads, which can attach colonies to stainless steel in a matter of hours. Once attached, the colonies are extraordinarily difficult to eliminate, so the best strategy is to control the colonies through treatment at the wells.

The bacterial counts were found to be relatively high and required some form of treatment. Further analysis indicated that the lime softening process, because of the mixing that occurred and the "sticky" consistency of the bacteria, did a relatively good job at removing the colonies. However, the proposed membrane softening process would not be as effective in this removal. The bacteria would foul the proposed membranes in the plant and could lead to breaching of the cartridge filters and lower plant efficiencies. Further review indicated the corrosion of the steel pipe at the lime softening plant could also be partially a result of the bacteria being brought in with the raw water.

To address this situation, the district proceeded with bids to initiate a routine disinfection program that added 2,000–6,000 ppm of calcium hypochlorite on a monthly basis. In addition, all ferrous material was removed from the wells, PVC slip lining was installed in the five original steel casings (with a loss of capacity), and a new composite column pipe was installed. All pumps were changed to bronze construction. These pumps were not only less costly but also faster to get and more resistant to microbiological attack than the stainless-steel pipe. All new wells were installed with PVC casings, bronze pumps, and a rubberized-composite column pipe.

The district monitored the total bacterial count changes and found that despite initial counts of more than 10,000 cfu/100 mL, with monthly disinfections, the counts began to decline relatively quickly. However, any prolonged time between disinfections or lowering the water table caused the bacterial counts again to climb. A side effect the district noted was that some well materials deteriorated during the disinfection program, including the composite column pipes, so they were removed from the wells each time they were chlorinated to address the deterioration of the column pipe.

Near the end of the construction of the membrane water treatment plant, the disinfection program lapsed when the contract ended. During plant startup, a significant quantity of the bacteria did get into the membrane units, requiring extensive cleaning of the membranes with bisulfite, citric acid, and hydrogen peroxide to eliminate the growth and restore the membrane efficiency. The extent of cleaning that was performed on these membranes is not desirable in a new membrane facility, where the membranes are expected to have a life of 5–7 years and the facility is expected to have a life of more than 30 years.

A second plugging problem occurred in well No. 12. Symptoms included a significant loss of specific capacity, inability to pump the required amounts, and inability to remove the pump. A television camera was dropped down the well. Significant quantities of iron-reducing bacteria (IRB), sulfur-reducing bacteria (SRB), and *Pseudomonas* bacteria were noted on the walls of the casing pipe. When the camera reached the pump, the casing appeared to have collapsed into the shape of a figure eight. The pump was below this collapsed point and could not be withdrawn.

Analysis of the problem indicated that the reduction of well capacity (due in part to the bacteria) and lower aquifer levels caused by drought in 1990 and 1991 caused the drawdown to reach a point just a few feet above the top of the pump. Because submersible pumps are designed to have several feet of water above them to keep them cool, when the water dropped to this point, the pumps heated enough to cause the PVC casing to buckle.

As a result of this analysis, all the submersible pumps were lowered 10–20 ft (3–6 m) depending on the well. The pump in well No. 12 was cut loose, and the possibility of cutting out the collapsed section of the pipe and splicing in a new one was considered. However, because of the cost for the splicing and the resulting loss of capacity, the well was abandoned.

The Collier County Water Sewer District's experience indicates that bacterial fouling can lead to column pipe damage, membrane fouling, collapsed casings, and lower pumping capabilities.

City of Venice Wellfield

The city of Venice is in Sarasota County, 40 miles south of Tampa, Fla. As a coastal community, the city has relatively little fresh water available. The overlying formation is undifferentiated sand and clay formation that lacks the ability to provide significant quantities of water. As a result, the city of Venice used the next aquifer formation, the Lower Hawthorn.

The production zone for the city of Venice is located between 200 and 320 ft (60–100 m) below sea level. Each well produces 1 MGD (3.8 MLD). The wells were installed in the 1980s. The raw water, although slightly brackish, has little color and hydrogen sulfide and thus lends itself to treatment with low-pressure reverse osmosis.

The city had a 4.0-MGD (15-MLD) low-pressure reverse-osmosis plant to treat the brackish Hawthorn water. Typically, low-pressure reverse-osmosis systems have a

70 percent recovery rate. However, because of high sulfate levels in the raw water, the city of Venice can recover only 50 percent of the raw water as permeate. As a result, nearly 8 MGD (30 MLD) is required to produce 4 MGD (15 MLD) for the distribution system.

The city experienced problems in several of its wells, including an increase in drawdown, a decrease in specific capacity, and pipe and pump cavitation. In addition, some sand was found in the prefilters. An increase in sulfates and chlorides was noted, and slime began to be produced on stainless steel column pipes in the pumps. A video camera inspection was made in several wells. In one well 227 ft (69 m) below sea level, sand and other materials were entrained in the open hole of the formation. The video indicated that the formation was producing intermittent fine sand and silt that were deleterious to the cartridge filters and the reverse-osmosis membranes. In addition, the wells contained a significant quantity of *Pseudomonas* bacteria species. The bacteria were attacking ferrous materials and causing the formation to plug, resulting in increased drawdown, pump failures, and lower water quality. The bacteria also have a potential deleterious effect on the reverse osmosis process efficiency.

In response to the analysis, the city instituted a routine disinfection program of 6,000 ppm of chlorine monthly. Over time, the city staff began to perform the routine disinfection, and the timing has decreased to once every 90 days. A program to routinely sample for microbiological parameters was instituted, and a water quality and water-level monitoring program was developed. All pumps were changed from stainless steel to bronze.

The sand problem could have been partially caused by the check valves opening and shutting on the wells, so slow-opening and -closing check valves were installed on each well to reduce the sand production. A sand separator was installed before the headworks of the pretreatment plant.

A second well showed similar signs: an increase in drawdown, bacterial counts, TDS, chlorides, and sulfates. Clogging also occurred. In viewing the well, the production well had become connected with an improperly abandoned well, constructed seven years earlier in a deeper saline zone of the Floridan aquifer. The abandoned Floridan well was contributing to poor water quality and bacteria via fissures located in the lower portion of the well.

As a result of the physical log and television survey, the city plugged the lower portion of the well with Type II sulfate-resistant cement between 350 and 450 ft (107–137 m) (Figure 4-7). The well was then disinfected, and routine monitoring of the bacteria and chlorides continued. Compilation of information after the plugging showed a steady and continuing decline of the chloride levels from more than 2,500 mg/L to about 500 mg/L over a period of just a few weeks (Figure 4-8).

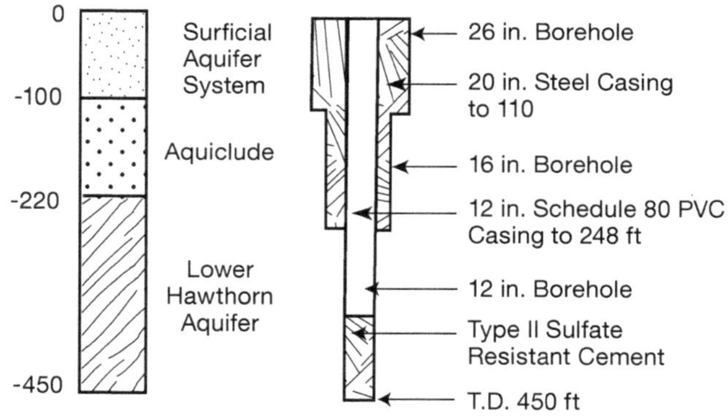

PVC: polyvinyl chloride, T.D.: total depth

Figure 4-7 Venice reverse-osmosis well construction after rehabilitation

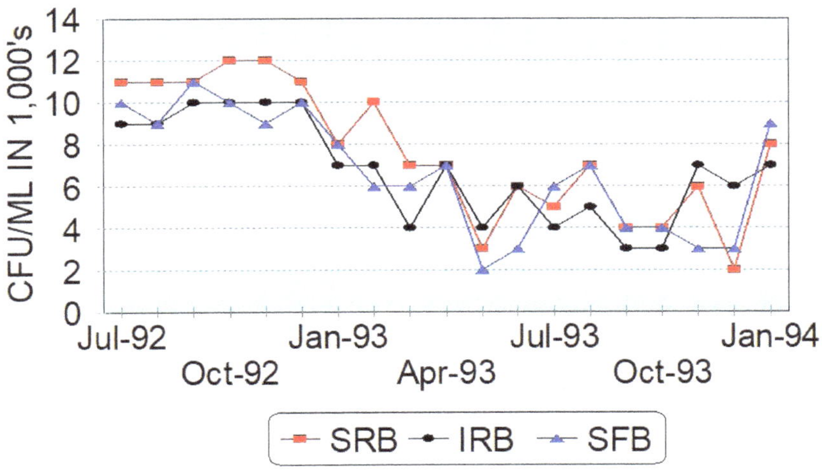

Source: Courtesy of Frederick Bloetscher

IRB: iron-reducing bacteria, SFB: segmented filamentous bacteria, SRB: sulfur-reducing bacteria

Figure 4-8 Changes in bacterial colonies after treatment began in December 1992

Conclusions from the Venice wellfield indicate that microbiological fouling may be a more prevalent problem than anticipated and that more sophisticated treatment methods may be more susceptible to microbiological fouling than more traditional methods. In addition, deeper wells may be severely affected by improperly abandoned wells. A full investigation of such wells must be conducted. The disinfection program restored the capacity of the wellfield, and the plugging of the well to seal off the effects of the improperly abandoned well did not significantly decrease the well's capacity.

Elkhart, Ind.

The South Wellfield in Elkhart, Ind.—one of three operated by the city's Department of Public Works and Utilities (DPWU)—is developed in the glaciofluvial outwash Yellow Creek tributary of the St. Joseph River aquifer. This wellfield is developed with three high-capacity screened "gravel-wall" wells and supplies a conventional aeration/pressure-filtration water treatment plant. Over time, these wells have experienced performance decline, adversely affecting the economy of the plant and its operations, with periodic attempts to restore production capacity.

Wells in the South Wellfield have experienced a decline in performance since at least 1971, when the first rehabilitation was conducted on well No. 1 (the northernmost of three). Each of the wells was treated several times. From the outset, the problem was attributed to "iron bacteria" and treated for such periodically. In its 1998 analysis, a biofouling cause was confirmed. A review of the treatment history since 1971 showed a pattern of continual decline despite repeated treatment. However, this decline was reversed somewhat by rehabilitation events. Specific capacity (yield per drawdown) is a readily calculated indicator of hydraulic performance change in wells.

Well No. 1, having the lowest initial specific capacity of the three (35 gpm/ft), declined below the optimal pumping economics point most quickly. Rehabilitation was first attempted six years after completion with no improvement, and then the well was permitted to decline in performance to uneconomical levels before a series of treatments from 1981 to 1989 kept specific capacity in the range of 20 to 30 gpm/ft. Treatment effectiveness then fell off rapidly, with specific capacity falling to as low as 2 gpm/ft, despite alternative methods of treatments, until the well was effectively abandoned in 1995. Table 4-3 summarizes treatments in well No. 1.

Wells No. 2 and 3, with higher initial specific capacities (51.2 and 88.6 gpm/ft, respectively), appeared to decline in performance more slowly. Well No. 2 was not rehabilitated until 21 years after original construction and specific capacity had fallen to 63 percent of original. Well No. 3 had a similar history but was permitted to drop to <40 percent of original capacity in 14 years.

On both wells No. 2 and 3, two rehabilitations each were performed in 1987–1991 once problems were recognized by the DPWU water staff. Chemicals used are as indicated. Table 4-4 summarizes the results.

Table 4-3 Treatments in well No. 1 summary

Date	Treatment	Before Capacity*	After Capacity
Dec. 1971	Acidization (A-6), phosphate (P-6, B-6), surging	28†	27.90
Sept. 1982	Acidization (A-6), phosphate (P-6, B-6) with HTH, surging‡	18.9	26
Sept. 1985	Phosphate (P-6) with HCl acidization and A-6, surging	20	28.3
Dec. 1987	P6 + HTH, light acidization, alternating, surging	26	26
Nov. 1989	Phosphate and acidization, chlorine and wetting agent, phosphate + wetting agent, surging	20.7	23.7
Oct. 1991	Phosphate with Cl_2, wetting agent, acidization alternating, surging	19.1	18.6
Mar. 1992	Surged and caustic soda added	12.5	10.65
Mar. 1993	Sonar-Jet® treatment	10	07
Mar. 1995	Aqua Freed®§ treatment	02	11

Source: Courtesy of Frederick Bloetscher

* Capacity = specific capacity (yield Q in gpm per drawdown in ft.)

† Original capacity = 34.6 gpm/ft.

‡ Treatments typically included alternating treatment chemical types and surging. Several hundred pounds of chemical was typically used.

§ Subsurface Technologies Inc., 40 Stone Castle Road, Rock Tavern, NY 12575

Table 4-4 Well No. 2 results

Events	Treatment	Before Capacity	After Capacity	Percentage of Original Capacity
	Well No. 2			
1987	Acidization, phosphate, surging	34	44	86
1991	Acidization, phosphate, surging	33	41*	80
	Well No. 3			
1987	Acidization, phosphate, surging	34	62.5	71
1991	Acidization, phosphate, surging	48	65.6†	74

Source: Courtesy of Frederick Bloetscher

* Capacity for 898 gpm.

† Capacity for 932 gpm.

For well No. 2, the 1987 treatment restored capacity to 86 percent of original, but capacity declined to below the 1986 precleaning value by 1991. The 1991 treatment restored capacity to 82 percent of original. However, the capacity reported was for 898 gpm (3,400 L/min), not 1,420 gpm (5,368 L/min). Specific capacity for any well at any point in time declines with increased pumping rate (Driscoll 1986). An estimated capacity for 1,420 gpm (5,368 L/min) at that time would have been less than 25 gpm/ft. Capacity then was permitted to decline precipitously to 23 gpm/ft at 800 gpm (3,025 L/min) before cleaning in September 1998.

For well No. 3, the 1987 treatment restored capacity to 71 percent of original, but performance declined to 53 percent of original by 1991. The 1991 treatment restored a reported capacity (for 932 gpm [3,523 L/min]) to 74 percent of original. An estimated capacity for 1,240 gpm (4,690 L/min) at that time would have been less than 49 gpm/ft. Capacity then was permitted to decline precipitously to <20 gpm/ft at 393 gpm (1,485 L/min) before cleaning in September 1998.

Contributing Factors in Well Performance Decline in the South Wellfield

An analysis of the history of treatment performance and well performance decline in these wells shows several contributing factors:

- The aquifer and well conditions have clogging potential. The working mechanisms are a combination of fine sediment migration from the glaciofluvial formation (mixed particle sizes) and biofouling. Fine sediment migrates toward the well. Biofouling theoretically forms in a cylindrical band through the depth into the formation where iron oxidizes to the screen face. Although biofouling does reduce hydraulic conductivity, it clogs more effectively because it traps in-migrating particles.
- The wells were permitted to decline in performance below the point at which full-performance recovery was possible. Below about 75–85 percent of original or target-specific capacity, a great amount of development energy is required to restore performance and especially to remove nutrients and residual debris to slow the return to well decline after cleaning.

Tragically, Elkhart's wellfield operations team from the mid-1980s to early 1990s had a well maintenance monitoring and treatment plan in place that could have halted decline earlier. However, this plan was permitted to lapse for several reasons. This kind of intermittent well maintenance history is more the exception than the rule in wellfield management.

Before 1998, phosphate-containing surfactant compounds were used in each treatment in large quantities. These were selected with the best of intentions on the basis of information provided by chemical suppliers and short-term (<10 year) experience in wellfields (including Elkhart's) that showed good initial results. However, phosphorus-containing surfactants are suspected of ultimately being counterproductive in well rehabilitation use because of residual phosphate (a limited nutrient in groundwater). Phosphorous is adsorbed to clays by cation exchange and available for bacteria to use in metabolism and cell growth and development (e.g., Smith 1995; Borch et al. 1993; Smith n.d.).

A condition commonly observed in sand-and-gravel wells treated repeatedly over time using phosphorus-containing compounds is a change in the type of biofouling present. It is transformed from a low-biomass filamentous form toward a bulkier, slimy type of biomass that is more difficult to remove using conventional rehabilitation methods. This change results in an acceleration of the performance decay in such wellfields. The change from short-term success to long-term acceleration of decline seems to be illustrated by the capacity history graph supplied by Peerless Midwest for well No. 1. Successes in the 1980s are followed by rapid declines in performance persisting to the present.

Evidence of a possible change in biofouling in the DPWU South Wellfield was provided by a review of color downhole videos performed on well No. 1. Whereas the problem was described in the past as "iron bacteria" (filamentous iron-related biofouling), recent videos showed a more gray, flocculent, slimy growth. BART methods and microscopy (methods per Standard Methods 2022; Smith 1996, 1992) confirmed the potential for intense slimy growth. Additionally, active denitrifying microflora were detected. These oxidize Fe^{+2} to Fe^{+3} anaerobically, opening up the possibility of a deep-set Fe^{+3} clog.

The effectiveness of conventional mechanical development used in past treatments was difficult to evaluate on the basis of file information, but the approach to treatment before 1998 was focused more on chemical application than on development action. Less-than-optimal redevelopment likely resulted in incomplete removal of clogging mass from the gravel pack and formation.

Because the wells appeared to be fundamentally sound and the cost of rehabilitation to restore performance was favorable compared with that of new construction, the consultant recommended rehabilitation over either well reconstruction or abandonment and new construction. Target yields and specific capacities were calculated on the basis of pumping goals (production needed and maximum drawdowns) and power efficiency (using Helweg et al. 1983 formulas).

On the basis of the analysis of causes, a Blended Chemical Heat Treatment (BCHT™) program (process documented in Leach et al. 1991; Smith 1995; Alford and Cullimore 1999) was recommended to break through the expected clogging material and restore performance. The BCHT process (which uses a mixture of chemicals, heated upon injection) has a history of effectiveness on difficult well clogs promoted by the slime-forming biofouling, similar to that detected in the South Wellfield tests.

In this case, the treatment comprised acetic acid (amended to reduce pH to <2) and nonphosphate polyelectrolyte,[§] jetted in at 180°F (82°C) at the nozzle, with a program of extensive mechanical development using double surge block and airlift pumping. This program was used on both wells No. 1 and 3.

Because of cost differences and as a comparison, well No. 2 was treated with hydrochloric acid, calcium hypochlorite, and development. Phosphate-containing compounds were not used in any treatments but replaced as surfactants by the nonphosphate polyelectrolyte products.

Well No. 1 was in extremely poor shape before cleaning (capacity = 8.2 gpm/ft at 402 gpm [1,520 L/min]). After the initial chemical charge, with minimal development, specific capacity fell to 5 gpm/ft. This was probably because of development action collapsing clogging material against the screen, but it resulted in some short-term handwringing. Surging and airlift began recovery over one week to 16.1 gpm/ft at 737 gpm (2,785 L/min), an economically viable level of performance for 1 million gpd, based on calculations. The effectiveness of development was hindered by a delay in commencement of development after chemical loading because of scheduling (under BCHT, development is most effective when commenced while the solution is still hot), some stoppage in development subsequently due to mechanical problems and process "choke points," and lack of initial effectiveness of development with the tools at hand.

Well No. 3 provided the most effective immediate response to the BCHT approach. After one chemical treatment pass and three days of development, capacity was restored to 55 gpm/ft at 770 gpm (2,910 L/min) from 15.6 gpm/ft at 686 gpm (2,593 L/min). Capacity reached 61.3 gpm/ft on July 23, 1998, when a large amount of silica sand was pumped in.

§ ARCCsperse CB-4 and PM-30, ARCC, Daytona Beach, Fla.

The screen was repaired, reducing capacity somewhat. Overall, performance was restored to somewhat less than 1987 post-treatment levels by the end of treatments in 1998.

Well No. 2 was treated differently, using hydrochloric acid, alternating with an alkaline (soda) and chlorine steps, with three days' development. Success in immediate redevelopment response here was also evident in increased specific capacity: from 22.8 gpm/ft at 800 gpm (3,025 L/min) to 38.7 gpm/ft at 1,002 gpm (3,788 L/min).

Comparing the effectiveness of the two chemical regimes will require evaluation over time. Acid-amended acetic acid has been shown in more than 3,000 well applications to perform better than hydrochloric acid and chlorine on very advanced slime-forming biofouling. However, in wells where the clogging is not compacted, as in well No. 2, various chemical treatments can have similar results. History with aggressive biofouling well environments shows that the benefits of both BCHT (and the amended acetic acid chemical choice) and effective redevelopment come with delayed decline in performance after rehabilitation, rather than in obvious immediate effects.

Long-term effectiveness of these treatments in the South Wellfield will depend on follow-up by the Elkhart DPWU. The following recommendations are being considered by the DPWU:

- An immediate short-term follow-up should be additional low-intensity redevelopment of each well in the South Wellfield in the next two years to complete the work begun with the 1998 rehabilitation actions. Each well should respond to additional development and light chemical treatment by increasing in performance if it is not permitted to decline in performance first.
- For further benefit, a program of professionally developed, city-administered maintenance evaluation and treatment is essential in the South Wellfield and, by extension, all three wellfields. A continued resumption in performance decline can be expected if no maintenance treatment actions are taken. The lapse in preventive maintenance treatments after 1992 almost certainly contributed to the state of the wells before the 1998 treatments.
- To best achieve these goals, all monitoring, treatment, and repair activities should be planned as part of a system-wide strategic wellfield maintenance program that is both systematic and effective.
- Within the maintenance plan, personnel training and wellfield equipment modification are recommended to make the process easier and more effective.
- A noncontractor advisory role was recommended for major treatment events. Professional assistance in this area by people highly experienced in well maintenance and rehabilitation helps to assure that a wellfield operator's objectives and best interests are served.

The Elkhart experience clearly shows what happens when wells are permitted to decline in performance. The experience in Eklhart's South Wellfield should not be considered unique. Prospects for success in well maintenance in other wellfields also depend on the kind of analysis, review, and planning documented here. As Elkhart has, any water supplier can benefit from honestly and completely scrutinizing successes (complete and incomplete), lapses, and failures in its maintenance history and taking advantage of the many improvements now available in the practice of well analysis, treatment, and maintenance.

SUMMARY

Although wells provide good service to most utilities for many years, wells are subject to fouling and other performance problems. These concerns include:

- mechanical failures, including failures of electrical motors and pumps and valves;
- poor and/or inconsistent operating and maintenance procedures;
- poor well development;
- poor well design and construction practices (including insufficient placement of grout; improper design of pumps, valves, and fittings; and excessive drawdown allowances);
- hydrogeologic constraints that are unassessed at the time of design or change over time, such as sand, clay, or rock layers that are unstable and collapse into the borehole; naturally occurring or induced fracturing and faulting; long-term water quality changes caused by changes to the hydraulic regime such as dams; water hammer to the aquifer; effects of mining the water or introducing chemicals and microorganisms; change in land use or recharge area; and naturally occurring phenomena (such as sinkholes, karst terrain features, or faults); and
- high silt or sand content caused by failing to develop the wells fully or intercepting sand or silt layers that have not or cannot be sealed off in the borehole or corrected in well design.

All these problems may exist in conjunction with—or because of—microbiological fouling problems in wells. As treatment technologies advance, the need to review and correct well performance problems, especially fouling concerns, has taken on greater significance.

REFERENCES

Ahmad, R., A. Amirtharajah, A. Al-Shawwa, and P.M. Huck. 1998. "Effects of Backwashing on Biological Filters." *Journal AWWA*. 90(12):62–73.

Alford, G. and D.R. Cullimore. 1999. *The Application of Heat and Chemicals in the Control of Biofouling Events in Wells.* Boca Raton, Fla.: CRC Press/Lewis Publishers.

Arihood, L.D. and D.A. Cohen. 1997. "Geohydrology and Simulated Ground-Water Flow in Northwestern Elkhart County, Indiana. Preliminary Water Resources Investigation Report." Indianapolis: US Geological Survey.

Block, J.C. 1989. *Viruses in Water Supplies—Detection and Identification*. New York: VCH Press.

Bloetscher, F., G.M. Witt, A.E. Dodd, and C.P. Dodd. 1998. "Prevention of Biofouling in Membrane Water Treatment Plants." American Water Works Association (AWWA) Annual Conference and Exposition. Dallas: June 1998.

Borch, M.A., S.A. Smith, and L.N. Noble. 1993. *Evaluation, Maintenance, and Restoration of Water Supply Wells*. Denver: AWWA Research Foundation.

Bouwer, H. 1991. "The Role of Groundwater Recharge in the Treatment and Storage of Wastewater for Reuse." *Water Science and Technology*. 24(9):259–302.

Chapelle, F.H. 1993. *Groundwater Microbiology and Geochemistry*. New York: John Wiley and Sons.

Cullimore, D.R. 1993. *Practical Ground Water Microbiology.* Boca Raton, Fla.: CRC Press/Lewis Publishers.

Cullimore, D.R. 2000. *Practical Atlas for Bacterial Identification,* Boca Raton, Fla.: CRC Press/Lewis Publishers.

Driscoll, F.G. 1986. *Groundwater and Wells.* St. Paul, Minn.: Johnson Division.

Gariboglio, M.A. and S.A. Smith. 1993. "Corrosión e Incrustación Microbiológia en Sistemas de Captación y Conducción de Agua: Aspectos Teóricos y Aplicados." Serie Investigaciones Aplicadas, Consejo Federal de Inversiones. Buenos Aires, Argentina.

Gerba, C.P. and G. Bitton. 1984. *Groundwater Pollution and Microbiology.* New York: John Wiley & Sons.

Goldrabe, J.C., R.S. Summers, R.J. Miltner. 1993. "Particle Removal and Head Loss Development in Biological Filters." *Journal AWWA.* 85(12):94ff.

Ground Water Science. 2000. "Evaluation of Problems with Closed Basin Division Salvage Wells, Rehabilitation Method Tests, Methods for Monitoring Well Deterioration, and Recommendations for Preventive Maintenance and Rehabilitation. A Comprehensive Report. Prepared for US Department of Interior." Denver: Bureau of Reclamation.

Krusemann, G.P. and N.A. de Ridder. 1994 (2nd ed.). "Analysis and Evaluation of Pumping Test Data, ILRI Publication 47." Wageningen, the Netherlands: International Institute for Land Reclamation and Improvement.

Helweg, O.J., V.H. Scott, and J.C. Scalmanini. 1983. *Improving Well and Pump Efficiency.* Denver: AWWA.

Imbrigiotta, T.E. and A. Martin, Jr. 1981. "Hydrologic and Chemical Evaluation of the Ground-Water Resources of Northwest Elkhart County, Indiana. Water Resources Investigations Report 81–53." Indianapolis: US Geological Survey.

Leach, R., A. Mikell, C. Richardson, and G. Alford. 1991. "Rehabilitation of Monitoring, Production and Recharge Wells." *Proc. 15th Annual Army Environmental R&D Symposium (1990).* CETHA-TS-CR-91077. Aberdeen Proving Ground, Md.: US Army Toxic and Hazardous Materials Agency.

Rose, J.B. and R.P. Carnahan. 1992. "Pathogen Removal by Full Scale Wastewater Treatment." Tampa, Fla.: University of South Florida.

Roscoe Moss Co. 1990. *Handbook of Groundwater Developmen*t. New York: Wiley-InterScience.

Sellwood, J. and J.V. Dadswell. 1991. "Human Viruses and Water." In *Current Topics in Clinical Virology* (P. Morgan-Capner, editor). Colindale, UK: Public Health Laboratory Service.

Smith, S. n.d. Personal communication (Layne Inc.)

Smith, S.A. 1992. *Methods for Monitoring Iron and Manganese Biofouling in Water Supply Wells.* Denver: AWWA Research Foundation.

Smith, S.A. 1995. *Monitoring and Remediation Wells: Problem Prevention, Maintenance, and Rehabilitation*. Boca Raton, Fla.: CRC Press/Lewis Publishers.

Smith, S.A. 1996. "Monitoring Biofouling in Source and Treated Waters: Status of Available Methods and Recommendations for Standard Guide." In *Sampling Environmental Media*, ASTM STP 1282 (J.H. Morgan, editor). West Conshohocken, Pa.: ASTM International.

Smith, S.A. and A.E. Comeskey. 2001. "Field Documentation of Improvements in Well Performance Using Innovative Rehabilitation Methods." In *Proc. of the World Water and Environmental Resources Conference*. Orlando, Fla.: ASCE Environmental & Water Resources Institute.

Smith, S.A. 2001. *Recent Innovations in Well Rehabilitation*. Upper Sandusky, Ohio: Ground Water Science.

Standard Methods for the Examination of Water and Wastewater. 2022 (24th ed.). APHA, AWWA, and WEF: Washington, DC.

Sustainable Water Well Infrastructure (SWWI). 2000. City of North Battleford, Well 15, 1997 Field Test of UAB Water Well Treatment Technology. Regina, SK, Canada: PFRA Technical Services Earth Science Division.

Wang, J.Z., R.S. Summers, and R.J. Miltner. 1995. "Biofiltration Performance: Part 1, Relationship to Biomass." *Journal AWWA*. 87(12):55ff.

Yates, M.V., C.P. Gerba, and L.M. Kelley. 1985. "Virus Resistance in Groundwater." *Applied and Environmental Microbiology*. 49(4):778–781.

Yates, M.V., S.R. Yates, J.M. Wagner, and C.P. Gerba. 1987. "Modeling Virus Survival and Transport in the Subsurface." *Journal of Contaminant Hydrology*. 1(3):329–345.

Chapter 5

Groundwater Quality and Treatment

This chapter provides an overview of common constituents in groundwater and common treatment techniques applied to groundwater including aeration, oxidation, softening and ion exchange, filtration, adsorption and absorption, corrosion control, disinfection, and fluoridation. When used in combination to achieve treatment goals, the interactions and effects of unit processes must be carefully considered.

Groundwater quality is as important as the quantity of water available, regardless of whether the source is located near the land surface or at great depths. In surficial aquifers, precipitation, which travels overland to streams or infiltrates below the land surface to become soil water and groundwater, dissolves rock minerals and organic matter in the process. It can also carry surficial contaminants into the formation. The latter is less likely with deep, confined aquifers, but they may have other challenges such as metals. As a result, groundwater contains a variety of chemicals resulting from natural sources, as well as from overlying land uses and local management practices.

CHEMICALS IN GROUNDWATER

The 14 principal chemical constituents of natural surface and groundwater and their effects are listed in Table 5-1. Only small to moderate amounts of these substances occur in most freshwater sources. Moderate amounts of dissolved minerals make water more palatable, as mineral-free water tastes flat to most people. Minerals are also important to human health and plant and animal growth. Some minerals reduce the corrosiveness of the water in pipelines and storage tanks.

The chemical, physical, biological, and radiological quality of groundwater varies widely. Acceptable quality depends on water use. For example, the criteria for safe and healthy drinking water are much more stringent than those for water used for industrial and agricultural purposes.

Chemical and Physical Characteristics

Groundwater's chemical and physical characteristics are determined by the characteristics of the physical and geochemical setting of the groundwater. The most common of these are hydrogen-ion concentration (pH), temperature, hardness, and gas content. These characteristics are discussed in the following paragraphs.

pH. Most groundwater has a pH value ranging from 6.0 to 8.5. Natural groundwater having a pH greater than 9.0 is unusual, except when contaminated. Many thermal springs yield water with a pH lower than 6. River water unaffected by contaminants generally has a pH between 6.5 and 8.5. Handheld devices are available to accurately measure pH in the field.

Temperature. In contrast to the seasonal and diurnal fluctuations of surface water temperature, in many cases the temperature of groundwater is fairly constant. The exception is the temperature of groundwater near the surface, which may fluctuate several degrees during the year in response to the seasons. A constant groundwater temperature helps maintain the palatability of drinking water.

The mean temperature of groundwater at shallow depths is generally 2–3°F (1°C) above the mean annual air temperature. Below this zone of solar influence, the temperature of groundwater increases at a rate of approximately 1°F (0.6°C) for each 64 ft (20 m) of depth. This increase mirrors the geothermal gradient of the earth's crust. With a limited number of exceptions, groundwater pumped from deep wells has higher temperatures than that pumped from shallow wells.

Table 5-1 The most representative natural chemical constituents in water, concentrations, and effects on the usability or limitations of the water

Constituent	Concentrations in Natural Water	Effects on Usability of Water
Silica (SiO_2)	Ranges generally from 1.0 to 3.0 mg/L, although as much as 100 mg/L is common; as much as 4,000 mg/L is found in brines.	In the presence of calcium and magnesium, silica forms a scale in boilers and on steam turbines that retards heat and fluid flow; the scale is difficult to remove. Silica may be added to soft water to inhibit corrosion of iron pipes.
Iron (Fe)	Groundwater having a pH of less than 8.0 may contain 10 mg/L; rarely, as much as 50 mg/L may occur. Acid water from thermal springs, mine wastes, and industrial wastes may contain more than 6,000 mg/L.	More than 0.1 mg/L precipitates after exposure to air; causes turbidity; stains plumbing fixtures, laundry, and cooking utensils; and imparts objectionable tastes and colors to food and drinks. More than 0.2 mg/L is objectionable for most industrial uses.
Manganese (Mn)	Generally 0.20 mg/L or less. Groundwater and acid mine water may contain more than 10 mg/L. Water at the bottom of a stratified reservoir may contain more than 150 mg/L.	More than 0.2 mg/L precipitates on oxidation; causes undesirable tastes, deposits on food during cooking; stains plumbing fixtures and laundry; and fosters growth in reservoirs, filters, and distribution systems. Most industrial users object to water containing more than 0.2 mg/L.

(continued)

Table 5-1 The most representative natural chemical constituents in water, concentrations, and effects on the usability or limitations of the water *(Continued)*

Constituent	Concentrations in Natural Water	Effects on Usability of Water
Calcium (Ca) Magnesium (Mg)	Calcium: averages about 15 mg/L in surface water, higher in groundwater. As much as 600 mg/L in some western streams; brines may contain as much as 75,000 mg/L. Magnesium: as much as several hundred milligrams per liter in some western streams. Ocean water contains more than 1,000 mg/L. Brines may contain as much as 57,000 mg/L.	Calcium and magnesium combine with bicarbonate, carbonate, sulfate, and silica to form heat-retarding, pipe-clogging scale in boilers and other heat-exchange equipment. Calcium and magnesium combine with ions of fatty acid in soaps to form soap suds; the more calcium and magnesium, the more soap required to form suds. A high concentration of magnesium has a laxative effect, especially on new users of the supply.
Sodium (Na) Potassium (K)	Sodium: as much as 1,000 mg/L in some western streams; about 10,000 mg/L in sea water; about 25,000 mg/L in brines. Potassium: generally less than about 10 mg/L; as much as 100 mg/L in hot springs; as much as 25,000 mg/L in brines.	More than 50 mg/L of sodium and potassium in the presence of suspended matter cause foaming, which accelerates scale formation and corrosion in boilers. Sodium and potassium carbonate in recirculating cooling water can cause deterioration of wood in cooling towers. More than 65 mg/L of sodium can cause problems in ice manufacture.
Carbonate (CO_3) Bicarbonate (HCO_3)	Carbonate: commonly 0 mg/L in surface water; commonly less than 10 mg/L in groundwater. Water high in sodium may contain as much as 50 mg/L of carbonate. Bicarbonate: commonly less than 500 mg/L; may exceed 1,000 mg/L in water highly charged with carbon dioxide.	Upon heating, bicarbonate is changed into steam, carbon dioxide, and carbonate. The carbonate combines with alkaline earths—principally calcium and magnesium—to form a crust-like scale of calcium and magnesium carbonate that retards flow of heat through pipe walls and restricts flow of fluids in pipes. Water containing large amounts of bicarbonate and alkalinity is undesirable in many industries.
Sulfate (SO_4)	Commonly less than 1,000 mg/L except in streams and wells influenced by acid mine drainage; as much as 200,000 mg/L in brines.	Sulfate combines with calcium to form an adherent, heat-retarding scale. More than 250 mg/L is objectionable in water in some industries. Water containing about 500 mg/L of sulfate tastes bitter; water containing about 1,000 mg/L may be cathartic.
Chloride (Cl)	Commonly less than 10 mg/L in humid regions; tidal streams contain increasing amounts of chloride (as much as 19,000 mg/L) as the bay or ocean is approached. About 19,300 mg/L in sea water; as much as 200,000 mg/L in brines.	Chloride in excess of 150 mg/L imparts a salty taste. Concentrations greatly in excess of 150 mg/L may cause physiological damage. Food processing industries usually require less than 250 mg/L. Some industries—textile processing, paper manufacturing, and synthetic rubber manufacturing—desire less than 100 mg/L.
Fluoride (F)	Concentrations generally do not exceed 10 mg/L in groundwater or 1.0 mg/L in surface water. Concentrations may be as much as 1,600 mg/L in brines.	Fluoride concentration between 0.6 and 1.7 mg/L in drinking water has a beneficial effect on the structure and resistance to decay of children's teeth. Fluoride in excess of 1.5 mg/L in some areas causes mottled enamel in children's teeth. Fluoride in excess of 6.0 mg/L causes pronounced mottling and disfiguration of teeth.

(continued)

Table 5-1 The most representative natural chemical constituents in water, concentrations, and effects on the usability or limitations of the water *(Continued)*

Constituent	Concentrations in Natural Water	Effects on Usability of Water
Nitrate (NO_3)	In surface water not subjected to pollution, concentration of nitrate may be as much as 5.0 mg/L but commonly is less than 1.0 mg/L. In groundwater, the concentration of nitrate may be as much as 1,000 mg/L where polluted but generally is less than 50 mg/L.	Water containing large amounts of nitrate (more than 100 mg/L) is bitter tasting and may cause physiological distress. Water from shallow wells containing more than 45 mg/L has been reported to cause methemoglobinemia in infants. Small amounts of nitrate help reduce cracking of high-pressure boiler steel.
Dissolved solids	The mineral constituents dissolved in water constitute the dissolved solids. Surface water commonly contains less than 3,000 mg/L; streams draining salt beds in arid regions may contain more than 15,000 mg/L. Groundwater commonly contains less than 5,000 mg/L, and most of it at shallow depths contains less than 1,000 mg/L; some brines contain as much as 300,000 mg/L.	More than 500 mg/L is undesirable for drinking and many industrial uses. Less than 300 mg/L is desirable for dyeing textiles and manufacturing plastics, pulp paper, and rayon. Dissolved solids cause foaming in steam boilers; the maximum permissible content decreases with increases in operating pressure.

Adapted from Durfor, C.N. and E. Becker 1964

Table 5-2 Water hardness scale

Hardness Range *mg/L of $CaCO_3$*	Description
000–600	Soft
061–120	Moderately hard
121–180	Hard
More than 180	Very hard

Hardness. Hardness is derived mainly from calcium and magnesium, although other polyvalent, positive cations may also contribute. These metallic ions inhibit shower and hand washing lathering by reacting with the soap to form undesirable precipitates and can combine with certain anions in boiler water to form efficiency-robbing scale on tank walls and in pipes.

In aquifers containing hard water, lowering the water level in a well during pumping and the corresponding reduction in water pressure at the intake screen may precipitate calcium and magnesium compounds that clog the well screen. In an unscreened well, the precipitates may clog the openings in the aquifer immediately adjacent to the well bore with comparable reduction in inflow of water.

Numerical scales for rating water hardness have been devised and published, including, for example, the scale in Table 5-2 (Durfor and Becker 1964).

The hardness of water used for domestic purposes is not objectionable in concentrations below about 100 mg/L. The hardness of groundwater throughout much of the United States is less than 100 mg/L. However, groundwater in gypsiferous and carbonate bedrock formations of the north-central region (including North Dakota, South Dakota, South Texas, Iowa, and parts of surrounding states) and groundwater in other parts of the nation that is underlain by sedimentary rocks rich in calcium and magnesium generally exceed this level. Limestone formations typically exceed this concentration as well. In these areas, hardness levels of 300 mg/L are common, and levels as high as 1,000 mg/L occur in some places.

Gases. As precipitation falls through the atmosphere, it encounters soluble gases that may combine with the water droplets. Dust and other particulate matter suspended in the air add chemical constituents to the water. The combination of precipitation and carbon dioxide (CO_2) forms carbonic acid (H_2CO_3), increasing the acidity of precipitation. Additional carbon dioxide, originating from organic processes at the land surface and in the soil zone, dissolves in groundwater, further increasing its acidity and capacity to dissolve mineral matter. Carbon dioxide is a byproduct of biological activity. Where the groundwater has high total organic carbon, carbon dioxide is often related to microbiological activity. Carbon dioxide also lowers the pH of the water.

Oxygen (O_2) and hydrogen sulfide (H_2S) are other important gases in groundwater. The concentration of dissolved oxygen in shallow groundwater is usually less than 10 mg/L, and in deep-lying groundwater, it may be virtually absent. Dissolved oxygen is harmless to health and may improve the palatability of water. Dissolved oxygen contributes to water's corrosiveness to metals, most aggressively where carbon dioxide or low pH is present. Hydrogen sulfide gas is generated in groundwater by decomposing natural organic substances and sulfate-reducing bacteria acting on organic materials. Hydrogen sulfide is corrosive in the gaseous state and combines with water to form a weak acid solution. Noticeable concentrations of hydrogen sulfide indicate active sulfur bacteria activity.

Methane (CH_4), generated by the decomposition of vegetation and other organic materials, is prevalent in groundwater and soil moisture zones in small concentrations. Larger quantities, sufficient for domestic or small industrial heating and energy requirements, may be formed in peat bogs, coal mines, or large landfills containing thick deposits of decomposing organic wastes where the methane is created by bacteria activity. Fires and explosions in mines, basements, water wells, and petroleum wells often are attributable to the accumulation of methane gas.

GROUNDWATER CONTAMINATION

Groundwater contamination is a widespread and challenging problem. Contaminants often flow undetected into aquifer formations, migrating through the aquifer until a sizable portion of the aquifer has become degraded. Even where physically and economically practical, rehabilitating a contaminated aquifer is difficult, complicated, time-consuming, and expensive. In arid regions, where water resources are limited, aquifer remediation may be the only option for a reliable, long-term water source. Ongoing monitoring of sentinel wells is the only means to monitor such contamination.

Irreversible damages to some of the nation's groundwater resources have stimulated efforts to reduce the influx of contaminants at their sources. Wellhead protection efforts developed as a part of the Safe Drinking Water Act (SDWA) are a result of exposure of surficial aquifers to contamination. Prevention is simpler, more effective, and less costly than cleanup measures. However, even after elimination of contamination sources, aquifers can remain contaminated for decades as a result of the slow rate of groundwater movement and the corresponding slow rate of dilution and flushing of contaminating substances. Biodegradation, the breakdown of contaminants by microorganisms in the subsurface, is commonly used today for cleaning up contamination in many areas.

Protecting water quality in wells requires wellhead protection efforts as required in the SDWA. For additional discussion regarding source water and wellhead protection, refer to ANSI/AWWA G300 (Source Water Protection) and M73 (*Groundwater Management*). Up-to-date information regarding drinking-water regulations can be found on the US Environmental Protection Agency (USEPA) website. Table 5-3 is current as of 2022. State and provincial standards may vary; therefore, it is the responsibility of utilities to understand requirements specific to their location.

Table 5-3 Federal USEPA Primary Drinking Water Regulations (NPDWRs)

Constituent	Maximum Concentration *mg/L unless specified*
Alpha (Total)	15 pCi/L
Arsenic	0.010
Barium	2
Benzene	0.005
Beta particle and photon activity	4 millirem/year*
Cadmium	0.005
Carbon tetrachloride	0.005
Chromium (total)	0.1
Copper	Treatment technique; action level = 1.3
2,4-D	0.07
p-Dichlorobenzene	0.075
1,2-Dichloroethane	0.005
1,1-Dichloroethylene	0.007
Endrin	0.002
Fluoride	4.0
Lead	Treatment technique; action level = 0.015
Lindane	0.0002
Mercury (inorganic)	0.002
Methoxychlor	0.04
Nitrate (measured as N)	10
Nitrite (measured as N)	1
Radium-226 and radium-228 (combined)	05 pCi/L*
Selenium	0.05
Total coliforms	<5.0% of monthly samples†
Total trihalomethanes	0.080
Toxaphene	0.003
2,4,5-TP	0.05
1,1,1-Trichloroethane	0.2
Trichloroethylene	0.005
Turbidity (surface water systems only)	1–5 NTU‡
Vinyl chloride	0.002

Source: USEPA NPDWRs
pCi/L: picocuries per liter
* Annual dose equivalent to the body or any internal organ.
† No more than 5.0 percent of samples are total coliform (TC)–positive in a month. (For water systems that collect fewer than 40 routine samples per month, no more than 1 sample can be TC-positive per month.) Every sample that has TC must be analyzed for either fecal coliforms or *E. coli*. If two consecutive TC-positive samples are found, and one is also positive for *E. coli* or fecal coliforms, the system has an acute MCL violation. (Source: USEPA NPDWRs)
‡ For systems that use conventional or direct filtration, at no time can turbidity (cloudiness of water) exceed 1 nephelometric turbidity unit (NTU), and samples for turbidity must be less than or equal to 0.3 NTU in at least 95 percent of the samples in any month. Systems that use filtration other than the conventional or direct filtration must follow state limits, which must include turbidity at no time exceeding 5 NTU. (Source: USEPA NPDWRs)

Inorganic Compounds

Nitrogen group. Nitrogen is a constituent of all proteins and is widely distributed in plants and animals. Major sources include:

- irrigated agriculture,
- dairy and livestock wastes,
- sanitary wastes (septic tanks in unsewered areas and wastewater treatment plant discharges),
- landfill leachate, and
- some manufacturing wastes that are disposed of in waste pits.

Three maximum contaminant levels (MCLs) have been established for nitrate, nitrite, and total nitrogen existing in the forms of nitrate and nitrite. Nitrate has long been regulated because of its acute human health effect of impairing the ability of blood to carry oxygen. For example, nitrogen has affected approximately thousands of drinking-water wells in California (see Figure 5-1, which shows regions where wells have exceeded the 10-mg/L nitrate standard).

Total dissolved solids (TDS). The predominant substances in municipal water supplies are inorganic minerals. Together, these minerals constitute TDS and commonly include sodium, calcium, magnesium, bicarbonate, sulfate, chloride, and silica (salt is the predominant constituent). A secondary MCL for TDS has been established because it is an important index of groundwater quality and usability. A high amount of TDS impairs aesthetics and practical uses of a municipal supply and is normally related to salt content, which cannot be removed by conventional processes.

Minerals. Minerals occur naturally in the earth's crust and dissolve in water. The minerals group has individual chemical MCLs for both major minerals and trace elements, all of which may reach elevated concentrations in groundwater through human activities or natural conditions. (Arsenic is an example that can be affected by both.) Major minerals include manganese, arsenic, sulfate, iron, chloride, and fluoride. Fluoride and arsenic are the only major minerals with a federal MCL. Major minerals dissolve readily in water and become concentrated in agricultural runoff and wastewater treatment plant discharges, as well as through evaporation of seawater and freshwater bodies. Sulfates and chlorides are regulated with secondary standards because of aesthetic considerations, the cathartic effect on humans, and the salty taste imparted to water. Fluoride is regulated because too high a level may cause mottled teeth enamel and osteosclerosis.

Trace elements usually include cadmium, chromium, barium, beryllium, copper, lead, selenium, mercury, aluminum, and silver. Trace elements in the minerals group with MCLs include silver, mercury, arsenic, and selenium. Trace elements can damage living organisms at low concentrations and tend to accumulate in the food chain. Trace elements have a wide variety of uses, including mercury in paints and batteries and cadmium in electroplating. Selenium, although an essential trace element in animal diets, is toxic at high concentrations.

Lead is a trace element that the USEPA has determined to be a health concern at certain levels of exposure, especially for children and pregnant women. Although lead-based paint is the major source of lead in the environment, lead can also come from the corrosion of household plumbing that contains lead pipes or copper pipes joined by lead solder. State and federal laws now require that only lead-free solder and other lead-free material be used when building or repairing plumbing systems. Major efforts are underway with legislation passing in 2022 to address lead and copper leaching in service lines. Aggressive groundwater will need to be treated to control corrosion, and lead services lines will need to be identified and removed going forward.

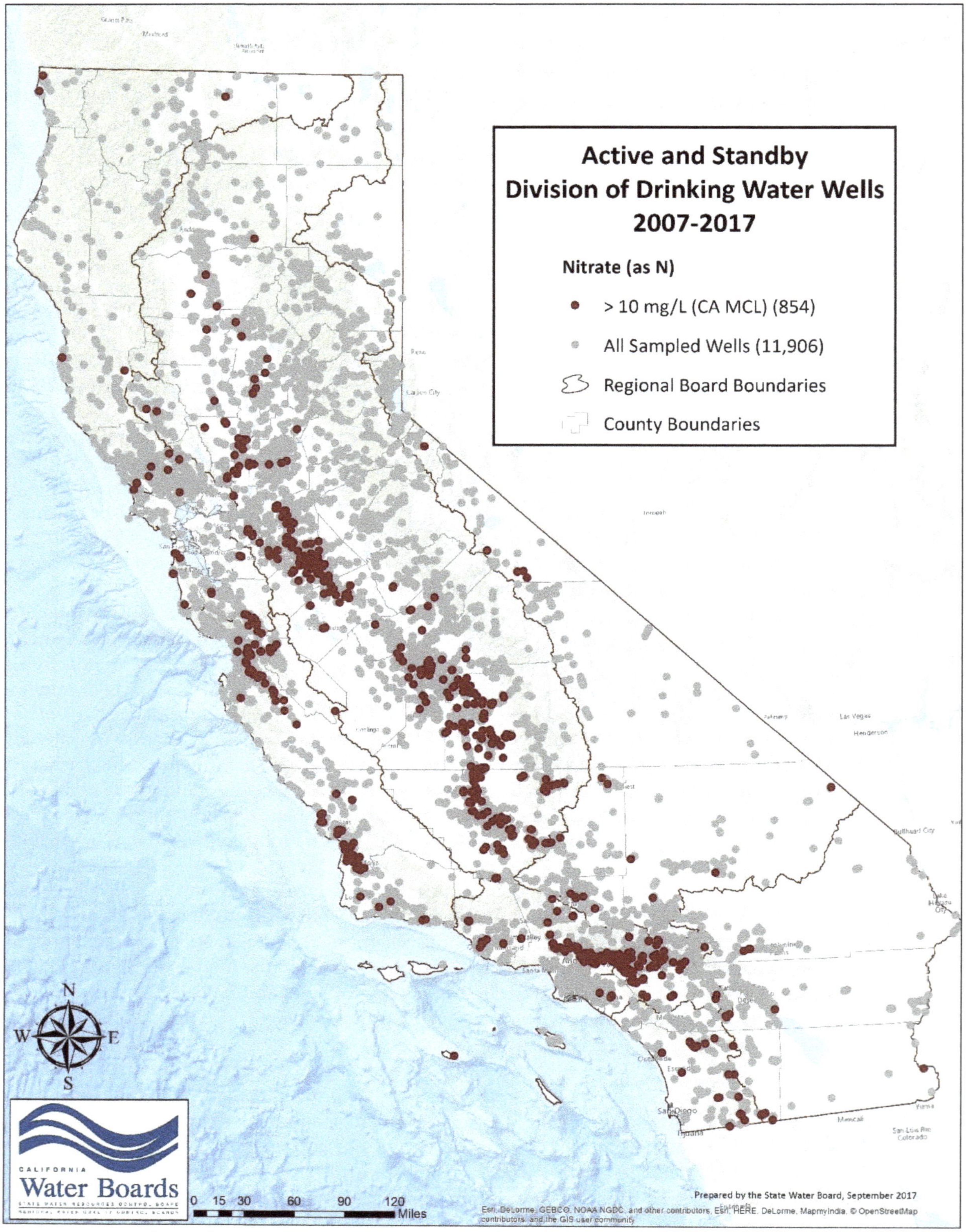

Source: California State Water Resources Control Board

Figure 5-1 Nitrogen in drinking water wells in California

Chromium is a naturally occurring element, the 11th most common in the earth's crust. Chromium is also used in many industrial processes, including electroplating, wood treatment, paints, and cooling tower treatment for corrosion control. The two most common species of chromium are chromium III, an essential dietary nutrient, and chromium VI, which can be toxic. According to the USEPA, background levels of chromium in US waters average 1 ppb, and drinking water averages 0.1 to 35 ppb. A USEPA survey of more than 3,800 US water taps found average chromium levels of 0.4–8 ppb, with varying amounts of chromium present. The USEPA has established an MCL of 100 ppb for total chromium. Although chromium VI is a human carcinogen when inhaled, scientific consensus has not been reached on health effects from ingestion.

The subject of recent regulatory focus, arsenic is a naturally occurring element that is present in both groundwater and surface water. It is the 20th most common element in the earth's crust (at average concentrations ranging from 1.5 to 5.0 mg/kg) and the 12th most common element in the human body. Arsenic can be naturally introduced to groundwater and surface water through erosion, dissolution, and weathering processes. Anthropogenic sources of arsenic include lumber, golf courses, agricultural practices, and general industry. Although concentrations of regulatory concern may be found in surface waters, arsenic is generally considered to be a groundwater issue. The toxicity of arsenic depends on its chemical form and the route and duration of exposure. Arsenic can produce both acute and chronic noncarcinogenic effects and is considered a carcinogen. The current MCL is 10 μg/L.

Radionuclides. Radionuclides are elements that spontaneously undergo radioactive decay and release energy in the process. Radionuclides include both human-made and naturally occurring isotopes. Several MCLs exist for radionuclides. For example, strontium-90 is a human-made radioactive isotope derived from fission products of nuclear reactor fuels and is present in fallout from nuclear bombs. Strontium-90 has a variety of uses, including industrial thickness gauges, static charge elimination, eye disease treatments, and cigarette density control. Uranium, a naturally occurring radioactive element, is used in nuclear reactors and in the production of nuclear weapons.

Radon-222 (radon) is a radioactive element generated naturally as a gas in the earth that dissolves in groundwater. It volatilizes during showers, bathing, and other activities, such as washing clothes. Radon spontaneously decays to radioactive daughter products and, in the process, changes from a gas to an ultrafine solid. Radon can be inhaled as well as ingested. Several studies have found a direct link between radon and human lung cancer.

Organic Groups

Volatile organic compounds (VOCs). VOCs have had widespread commercial and industrial use over the past 30 years. Industrial parts-cleaning and dry-cleaning operations are the top two users of VOCs, followed by manufacturers of chemical intermediates, electronics, pharmaceuticals, and textiles. Facilities using VOCs range from small dry cleaners to major aerospace and defense industries. Common sources of VOC releases include drains, pipelines, and discharges diverted to soil or aquifers and leaking underground storage tanks (LUSTs). VOC disposal and subsequent movement through landfills can increase the mobility of other toxic chemicals, all of which are ultimately reflected in the leachate contamination of groundwater.

Common solvent usage has included:

- trichloroethylene (TCE) for industrial parts-cleaning,
- tetrachloroethylene (PCE) for dry cleaning,
- carbon tetrachloride (CCl_4), formerly used for dry cleaning and fire extinguishers,

- 1,2-dichloroethane (DCA) in soaps and organic synthesis, and
- 1,1,2,2-tetrachlorethane for paint removers and in bleach manufacturing.

The physical properties and unreactive nature of VOCs that make them so useful also help make them persistent and mobile in groundwater. Their general toxicity to living organisms makes some VOCs resistant to biodegradation in the subsurface and a health issue for municipal water supplies (Montgomery 1996).

Fuels contain many constituents such as benzene, xylene, toluene, and ethylene. Finding any of those constituents suggests a fuel leak. However, various types of fuel (diesel, kerosene, gasoline) have different organic compounds, so testing for all may be costly. Using a surrogate has the benefits of lesser costs. For fuels, methyl tertiary butyl ether (MTBE) is one. MTBE is a fuel additive first introduced in the 1970s as an antiknock compound when lead was phased out of gasoline, so it is present in all vehicle fuel. In the 1990s, it was added to reformulated gasoline as an oxygenate to reduce smog production. If found, MTBE in groundwater suggests leaking underground and aboveground fuel tanks, pipelines and associated booster stations, refineries, and spills. Additional sources that may affect groundwater resources include surface water recreational activities using two-cycle engines. The removal of fuels from groundwater is a time-consuming and expensive process.

Pesticides. Pesticides are substances used to destroy or inhibit the action of plants or animal pests and include insecticides, herbicides, rodenticides, and nematocides. They are associated with irrigated agriculture, dairy, and livestock activities. Virtually all are toxic to humans to some degree, and they vary in biodegradability. Table 5-4 lists some of the regulated pesticides that have been detected in groundwater (Anderson 1990; National Toxicology Program 1999).

Perfluoroalkyl and polyfluoroalkyl substances (PFAS)/perfluorooctanesulfonic acid (PFOS). Among the most pressing current water quality issues is the presence of a group of chemicals termed PFAS and PFOS, which were discovered in source waters including groundwater. These substances have been in commercial use since 1940 because of their resistance to heat, water, and oil. PFAS and PFOS were initially used during World War II as a waterproofing coating on Allied tanks; later, they became the basis for polytetrafluoroethylene (PTFE)—used as a nonstick coating for pots and pans. Developed by

Table 5-4 Regulated pesticides detected in groundwater

Pesticides	Type	Additional Uses/Comments
Atrazine	Herbicide	Plant growth regulator; used for highway weed control
Bentazon	Herbicide	Food crops
Chlordane	Insecticide	Fumigant
2,4-D	Herbicide	Defoliant; agriculture and pasture weed killer; fruit drop control
1,2-Dibromo-3-chloropropane	Nematocide	Soil fumigant
Endrin	Insecticide	Banned US use and manufacture
Heptachlor	Insecticide	Banned except for termite control
Heptachlor epoxide	Insecticide	Heptachlor and chlordane degradation product
Lindane	Insecticide	Livestock, crops, lumber
Methoxychlor	Insecticide	Acaricide: livestock, dairy farms, food crops
Simazine	Herbicide	Algicide: agriculture, aquatic sites
2,4,5-TP	Herbicide	Banned plant growth regulator
Toxaphene	Insecticide	Not recommended for dairy activities

DuPont and 3M, these compounds are common in household cookware, lubricants, and the like. They have a history of various negative public health effects including cancer, and because of their prevalence in everyday items, everyone likely has these compounds in their bloodstream.

Over the years, these materials have found their way into groundwater near airports, factories, and military bases. While several states attempted to create regulations for these compounds, in March 2021 the USEPA published Regulatory Determinations for Contaminants on the Fourth Contaminant Candidate List,* which included a final determination to regulate two PFAS constituents in drinking water—perfluorooctanoic acid (PFOA) and PFOS. After receiving public comment, the USEPA finalized the rule in April 2024 with a maximum contaminant level (MCL) of 4 ppt for both PFOS and PFOA. This MCL is far lower than any current regulation and is close to the method detection limit (MDL) for these substances.

Groundwater utilities should monitor for these compounds because they may indicate surface contamination from industrial sites. Removal of PFAS and PFOS compounds is also being studied. Treatment or abandonment of contaminated wells are solutions. Utilities will receive more scrutiny of the issue especially in rural areas where oversight in the past may have been lacking.

Pharmaceutically active substances. It has been estimated that 70 percent of pharmaceuticals consumed pass through the body unchanged. Further, because concentrations of these substances are typically less than 1 ng/L, the microorganisms in wastewater treatment facilities are not induced to metabolize these substances as an energy source. As a result, current research indicates that many of these substances survive the biodegradation process and are discharged into receiving waters. Those that are altered can revert to their original form in the environment (Daughton and Ternes 1999). Recent research also indicates the presence of chemicals in water and wastewater that may disrupt the endocrine system of many species, including humans (Bloetscher and Plummer 2011). Endocrine chemicals are used by organisms to regulate important metabolic activities, such as ion balance, reproduction, basal metabolism, and fight or flight responses through changes in hormones secreted by the thyroid, parathyroid, pituitary, adrenal, sex, and other glands. Because endocrine systems are interconnected, effects on one will affect others as well. Chemicals—whether derived from pharmaceuticals, industrial emissions, or natural sources that interfere with endocrine systems of humans and wildlife—are termed endocrine disruptors, and those that elicit a pharmaceutical response in humans are termed pharmaceutically active substances (PAS). Research has identified more than 60 PAS that affect the endocrine system of animals and humans in nanograms per liter or lower concentrations in the ecosystem.

The first regulations requiring ecotoxicity testing for the registration of pharmaceuticals were established in Germany in 1995. As a result, most early research involving the identification and characterization of PASs in wastewater effluents and receiving waters occurred in Germany. In the United States, the 1996 SDWA Amendments and the Food Quality Protection Act of 1996 mandate comprehensive screening for estrogenic and antiestrogenic chemicals. Research initiatives by the AWWA Research Foundation and the Water Environment Federation Research Foundation address the problem. The first major survey published in the United States was in 2003.

Effects of endocrine disruptors in the environment have been observed. Both natural and synthetic chemicals may have disrupting effects; most observations involve species feminization and have been attributed to estrogenic compounds found in wastewater effluents (Lutz and Kloas 1999). The reverse also occasionally occurs, as in North Florida, where wastewater effluent from a paper mill is suspected in the masculinization of fish through

---

* www.epa.gov/ccl/regulatory-determination-4

the development of androgenic compounds in the process (Raloff 2001). In both cases, the sex change effect results in radically reduced resident fish populations, sexually shifted remaining populations, and potential loss of sustainability of the resident population. Also in Florida, alligator populations have been found to have greatly reduced fertility, traced to feminization and lack of development of reproductive organs in the male (Guillette et al. 1994). Nationally, many species have reportedly been affected (Colburn et al. 1997).

Until recently, the problem of PAS in the environment was not noticed because of the low concentrations and difficulty in tracing the compounds. Tracing drug residues is problematic because many potential endocrine disrupting chemicals have little in common structurally or in terms of chemical properties (DePledge and Billinghurst 1999). In addition, lists of active ingredients in pharmaceutical products are often not made available because of patent limitations, hindering the development of spectral signatures needed for analysis by gas chromatography/mass spectroscopy (Daughton and Ternes 1999). Furthermore, current effluent toxicity screening tests are not designed to detect endocrine disrupting and other effects of PASs, the effects of chronic exposure, or prenatal effects realized in offspring.

PAS and their ecological effects can be categorized (Raloff 2001; Daughton and Ternes 1999; Hirsch et al. 1999; Buser 1999; Ternes 1998) as shown in Table 5-5.

These chemicals are not reduced by wastewater organisms, creating a potential challenge where wastewater is being viewed as a water supply option, such as in certain areas of Florida, Nevada, and Arizona. It already has served this purpose at two sites in Texas and multiple sites in California. The largest project for injection of highly treated wastewater into the aquifer for recharge is currently practiced in Orange County, Calif – the Groundwater Replenishment System. The same concept processed (with RO and UV treatment) has been tested among several Florida communities for future potable use.

Table 5-5 Categories of pharmaceutically active substances (PAS)

Substance	Uses	Quantity	Effects
Estrogenic compounds	Contraceptive	1–5 μg/L	Feminization
Steroids (non-estrogens—androgen, testosterone, etc.)	Muscle development, various	Above 1 μg/L	Masculinization
Antibiotics	Reduce bacterial infection	Varies	Resistant pathogens
Blood lipid regulators	Cholesterol control	Up to 0.165 μg/L	Unknown
Nonlipid analgesics	Anti-inflammatory	0.5–1 μg/L	Unknown
Beta-blockers	Anti-anxiety	0.2 μg/L	Stimulate reproduction
Antidepressants	Increase serotonin, control behavior (Prozac, Ritalin)	Varies	Stimulate reproduction
Antiepileptics	Epilepsy control	Up to 6.3 μg/L	Unknown
Antineoplastics	Chemotherapy	0.017 μg/L	Toxicity, birth defects
Impotence drugs	Erectile dysfunction, blood stimulant	Unknown	Unknown
Retinoids	Skin diseases, antiaging, cancer	Unknown	Birth deformities
Contrast media chemicals	X-rays, CAT scans, diagnostics	15 μg/L	None
Fragrances and musks	Perfumes, colognes	Up to 0.4 μg/L	Toxicity
Preservatives	Antimicrobial	Unknown	Feminization
Disinfectants	Bactericides	0.05–0.15 μg/L	
Herbal remedies	Various	Varies	Various
Sunscreens	Protect skin from UV light	Unknown	Unknown

CAT: computerized axial tomography, UV: ultraviolet

The City of Pembroke Pines undertook an extensive investigation of alternative water supply options with specific pilot testing of aquifer recharge membranes. The city tested reverse-osmosis membranes to determine their ability to remove constituents, especially phosphorous and nitrogen compounds, as well as unregulated emerging substances. The membranes performed under all circumstances, and three log-removal goals were met or exceeded (Hart et al. 2011). Among the findings was that indirect potable reuse in Southeast Florida was found to be much more cost-effective than initially expected when consideration is given to actual environmental benefit (Bloetscher et al. 2014). The analysis also shows that the cost for reverse osmosis to accomplish aquifer injection may be far more competitive when compared with land application reuse systems than utilities may realize. Among the tests was a spike test to determine the removal capability of PAS. This study found that the use of low-pressure brackish water membranes along with ultraviolet (UV) light could achieve 3-log reduction, but neither treatment methods was fully successful without the other (Bloetscher et al. 2011; Hart et al. 2011). The findings indicated that the presence of PAS creates a treatment challenge not currently confronted by most groundwater systems.

Reusing water for potable use is not new. Many communities recharge their aquifers with irrigation water, reclaimed water (Hampton Roads) and by discharging to rivers and streams that are the water supplies for downstream users. The difference with the California, Florida, and Texas projects is that the intent was to take the water to potable supply wells or the head of the plant. A recent study by Bloetscher et al. 2024 concluded that direct potable reuse (RO- and UV-treated wastewater directed to the headworks for a water treatment facility) had a similar public health risk to Class I injection wells and at least one magnitude less than any other option.

Sources of Chemical Contamination

Potential contaminants are generated by virtually all of society's industrial, agricultural, urban, and rural activities. The principal sources of contaminating substances are shown in Table 5-6.

Contaminants may enter groundwater reservoirs by intent and design, such as deliberate placement in the subsurface through a waste-injection well. Also, contaminants may enter a groundwater system inadvertently—e.g., by leakage from a ruptured pipeline, as leachate from an inadequately sealed landfill, or because of agricultural fertilizer application.

Figure 5-2 illustrates the downward movement of a contaminant from a land-surface source (in this example, a waste lagoon) through the zone of aeration (the zone of rock and soil above the water table is unsaturated with water) and then into the aquifer. As illustrated, the contaminating liquid migrates downward and laterally into the aquifer. The recharge mound built up beneath the source of the contaminant may propel some of the

Table 5-6 Major sources of groundwater contamination

Point Sources	Non–Point Sources
Landfills	Agriculture
Superfund-type sites	Dairies and feedlots
Leaking underground storage tanks	Seawater intrusion
Wastewater plants	Urban runoff
Oil production and refining facilities	Oil, sewer, and other pipeline networks
Industrial and manufacturing facilities	Oilfield brine injection
Septic tanks	Atmospheric deposition
Spills and accidents	Acid mine drainage

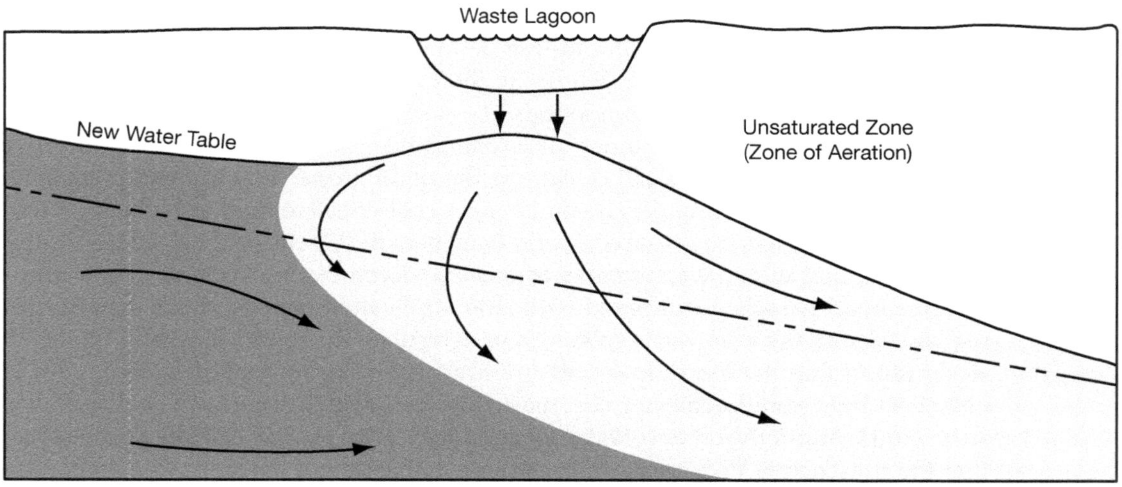

Figure 5-2 Flow of contamination from a ponded surface source into an aquifer

liquid in an upward direction (to the left in the illustration) for relatively short distances, but the dominant direction of movement for contaminant liquids with density similar to that of the groundwater is downward in the direction of general groundwater flow, indicated by the direction of slope of the old water table (to the right in the illustration).

The relative densities of incoming fluids and of the receiving water in the aquifer influence the pattern of contaminant movement. High-density organics—such as chloroform, DCA, and PCE—tend to "sink" to the lower part of the aquifer and are commonly called dense, nonaqueous phase liquids. Comparatively, fluids that are less dense than water, such as gasoline and oil, tend to "float" on the groundwater and are called light nonaqueous phase liquids, or L-NAPLs.

TREATMENT

Aeration

Aeration mixes water with air to transfer gas from the water to the air. Aeration is often used to:

- remove objectionable dissolved gases such as hydrogen sulfide and carbon dioxide,
- remove certain VOCs, and
- oxidize reduced constituents in the water such as iron and manganese (see Oxidation).

There are a few gases that are easily liberated from groundwater: carbon dioxide, hydrogen sulfide, and methane. Carbon dioxide removal increases the pH of the water. Aeration of water with significant amounts of hydrogen sulfide can create odor problems and a highly corrosive environment for treatment plants and equipment. Hydrogen sulfide is also toxic at certain concentrations that may not be detectable by onsite personnel (high concentrations burn out nasal sensors). The introduction of air into water containing microbiological populations can increase biological activity, potentially affecting downstream water quality. Aeration should not precede membrane filtration.

Figures 5-3, 5-4, and 5-5 show typical aeration or air-stripping devices. Natural draft aeration uses a device open to the atmosphere. Water enters the top, falls in or around trays, and is collected at the bottom. The turbulence and mixing of the cascading water

Source: Courtesy of Frederick Bloetscher

Figure 5-3 Tray aeration—aluminum

provide gas transfer. Figures 5-3 and 5-4 are tray aerators, which rely on the breeze and cascading water to make Henry's Law work to remove the gases. The aluminum in Figure 5-3 is covered in bacteria colonies, iron, and algae. Iron is also easily removed by stripping. Figure 5-4 shows how easily iron is removed. More recalcitrant gases require higher volumes of air to remove the dissolved gases. Forced or induced draft aeration uses a device similar to the natural draft aeration but is equipped with a blower. The blower forces air from the bottom of the aerator out the top. This countercurrent effect provides a greater level of gas transfer than the natural draft aerator and a higher air-to-water ratio.

If the draft aeration does not work efficiently enough, these gases can be removed with packed towers (see Figure 5-5). Packed towers have water sprayed down from the top and large volumes of air blown upward to facilitate the stripping process. The packed-tower concept is shown in Figure 5-6. Note the packing is a series of small objects designed to cause the water to break apart even further than the spray droplets. The blown air creates countercurrent movement of air for stripping. Because of its design, the packed-tower aerator can achieve high levels of efficient mixing and transfer and is typically used for treating VOCs. In many cases, the offgas containing the VOCs is captured for further treatment.

Source: Courtesy of Frederick Bloetscher

Figure 5-4 Tray aerator—plastic showing major iron deposition

Oxidation

In oxidation, one or more reactants lose or donate electrons (oxidize), and one or more reactants are reduced (gain or accept electrons). Oxidation reactions are very common for the treatment and disinfection of groundwater.

For example, soluble iron (Fe^{+2}) and manganese (Mn^{+2}) generally oxidize to ferric (Fe^{+3}) and manganic (Mn^{+4}) forms, which are insoluble and precipitate. Oxidation can remove color, tastes, and odors to varying degrees. Common chemical additives capable of providing oxidation include ozone (O_3), permanganate (MnO_{4-}), and chlorine (Cl^-). Although aeration can provide a level of oxidation, its effect is weak compared with direct application of the chemical oxidants.

Chlorine and chlorine compounds. Aqueous chlorine is one of the most effective chemical oxidants for use in manganese oxidation, carbon removal, and control of taste and odors. Unfortunately, free residual chlorine may combine with organic compounds to create trihalomethanes (THMs) and other byproducts. Chlorine can be used for oxidation.

Potassium permanganate. Potassium permanganate ($KMnO_4$) can oxidize most organic and many inorganic pollutants in water. Thermodynamic reactions form manganese dioxide (MnO_2), lending the remaining oxygen pair for use to oxidize iron and other contaminants.

Source: Courtesy of Frederick Bloetscher

Figure 5-5 Packed tower

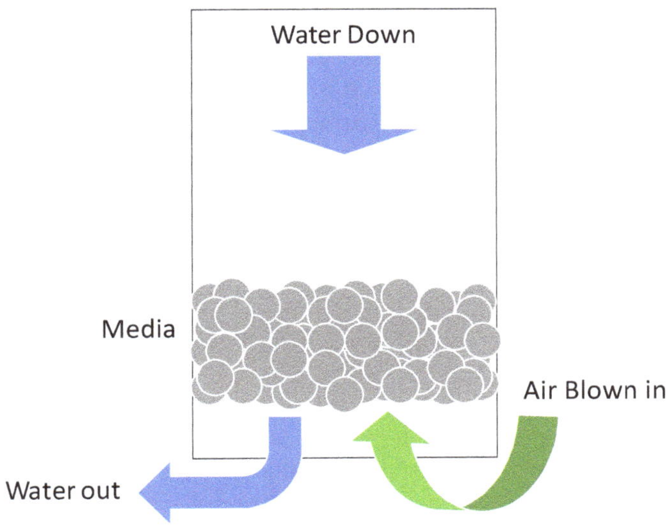

Source: Courtesy of Frederick Bloetscher

Figure 5-6 Interior of how packed tower works

Potassium permanganate is used to oxidate iron, manganese, cyanide, and phenols. It is used for taste and odor control, and for color removal. Often, potassium permanganate is injected into the water stream before a contact or oxidation tank. The tank provides sufficient detention time for the thermodynamic reactions to occur. Typical doses of permanganate are 1–2 mg/L. Permanganate is purple in color, and the color does not dissipate. As a result, the potassium permanganate dosage must be carefully determined and controlled, or the treated water will contain an excess of manganese and be pinkish-purple in color

Ozone. In aqueous systems, ozone reacts directly with contaminants, or indirectly, when ozone decomposes to form the hydroxyl radical. The indirect process is advanced oxidation. As a result of the hydroxyl radical, ozone is a powerful oxidant often used as pretreatment or as an intermediate treatment process. Ozone is useful for treating taste, odor, and color compounds because the substances in natural waters that cause tastes and odors often contain ozone-sensitive functional groups or unsaturated molecular bonds that can be broken by the hydroxyl radical. The organic compounds may not be completely oxidized and may change in molecular structure, which may require further treatment. Ozonation byproducts are discussed in the disinfection section of this chapter.

Ozone oxidizes synthetic organic chemicals (SOCs), an application favoring the high selectivity of molecular ozone. Ozone is most useful as an oxidant for phenolic pollutants and some pesticides with vulnerable functional groups.

Ozone must be generated onsite and used. The life span of ozone is about 30 min. Ozone typically bubbles up from the bottom of a tank that is 18–20 ft (5-6 m) deep. The excess gas must be recaptured because of its potential for corrosion and pollution. Ozone is expensive to operate, and the oxygen residual may complicate downstream processes (e.g., membranes).

Advanced oxidation. Advanced oxidation processes generate an oxidizing agent called the hydroxyl ($OH^{\cdot}$) radical, which is extremely reactive because it contains a single, unpaired electron. Unlike molecular ozone, the hydroxyl radical is not selective as an oxidizing agent. Advanced oxidation most commonly involves the use of three types of processes: hydrogen peroxide and ozone, UV light and ozone, and UV light and hydrogen peroxide. The UV light or hydrogen peroxide acts to encourage ozone decomposition to the hydroxyl radical in quantities sufficient to treat otherwise ozone-resistant contaminants. Most wastewater processes for indirect potable reuse or groundwater recharge use peroxide and UV lights for disinfection.

Advanced oxidation has been used to treat taste and odor compounds. Typically, the ozone dosage required is low. Laboratory and pilot plant studies have been conducted for the treatment of SOCs. Molecules that are refractory to molecular ozone (e.g., halogenated alkanes and alkenes) and some aromatics (e.g., benzenes) are reactive with the $OH^{\cdot}$ radical. Pilot-scale work indicates that advanced oxidation may be useful in removing THM precursors resistant to molecular ozone, such as certain ketones. As is the case with ozone, the formation of bromate and assimilable organic compounds (AOCs) is a concern with advanced oxidation. UV and hydrogen peroxide used after membranes have been demonstrated to remove pharmaceutically active compounds (Hart et al. 2011)

Ion Exchange

Hardness in drinking water increases the amount of soap needed to produce a foam or lather and causes scale in hot water pipes, heaters, and boilers. Hardness is caused by a high concentration of divalent metallic cations. The ion-exchange process consists of passing the water to be treated through a tank containing the ion-exchange resin. In the tank, positively charged ions are replaced with the more soluble sodium ions. Examples include calcium, magnesium, arsenic, iron, manganese, barium, strontium, and zinc. Ion exchange is also applied to the removal of anions such as nitrates. In this case, the anions (chlorides) replace OH^- ions. All ion-exchange filters must be backwashed and regenerated on a regular basis. The backwash is typically a salt-brined solution. The key is finding the correct resin to remove the constituent of concern.

Metals in drinking water are derived largely from contact with soil and rock formations. In general, hard waters originate in areas where the topsoil is thick and limestone formations are present. In other geologic formations, such as granitic-based materials, the groundwater contains less calcium and magnesium but may still contain too much iron and manganese. Although not considered to be health hazards, iron and manganese can impart color to water and can stain laundry and plumbing fixtures (generating "red water" complaints). Iron and manganese can precipitate in pipes and fittings and can encourage bacterial slime growths in hot water pipes, heaters, and boilers. Both are easily removed with ion exchange. Many metals, particularly arsenic, are also removed with specific ion-exchange resins. Arsenic may occur where pyrite crystals are found in limestone or dolomite.

Historically, this process was called sodium zeolite softening, because natural (e.g., greensand or glauconite) or synthetic zeolite was used as an ion-exchange material. The concept is that the ion-exchange resins are a catalyst, and the water to be treated is run through a bed of the resins, which removes the calcium or other constituent. An example of an ion-exchange reactor is shown in Figure 5-7.

Source: Courtesy of Frederick Bloetscher

Figure 5-7 Ion-exchange tanks

Modern ion-exchange materials consist of a matrix or hydrocarbon network such as polystyrene, which is copolymerized with divinylbenzene (DVB). The matrix is converted to a strong-acid cation exchanger or to a strong-base anion exchanger, depending on the type of ionizable groups attached to the network. In softening by ion exchange, only strong-acid, cation-exchange resins operating in the sodium cycle are used as exchange material.

Another version of ion exchange involves magnetic ion exchange (MIEX). In MIEX processes, the water is mixed in a mixing basin to create the exchange (see Figure 5-8). Otherwise, the process is the same.

Note: Ion exchange is suggested as one treatment option for long-chain PAS compounds, but the media cannot be recharged if PFAS are present.

Lime Softening Treatment

Cold lime softening is one of the oldest methods of water treatment. Lime (calcium oxide) is added to the water or lime in combination with soda ash (sodium carbonate). Lime softening is worthy of some discussion because it is distinctly a groundwater treatment solution and misunderstood to a degree. Lime softening relies on pH to work. The concept is to use the alkalinity diagram (Figure 5-9) to move the alkalinity species from bicarbonate to carbonate, which settles. The alkalinity diagram shows that at neutral pH, bicarbonate is the predominant species. Lime is added on the basis of the milliequivalents of polyvalent positive ions (calcium, magnesium, iron, etc.) in the raw water. Hence, raw water quality is necessary to dose properly. And pH must trend toward 10 to work effectively. Too often,

Source: Courtesy of Frederick Bloetscher

Figure 5-8 Magnetic ion exchange (MIEX) unit top

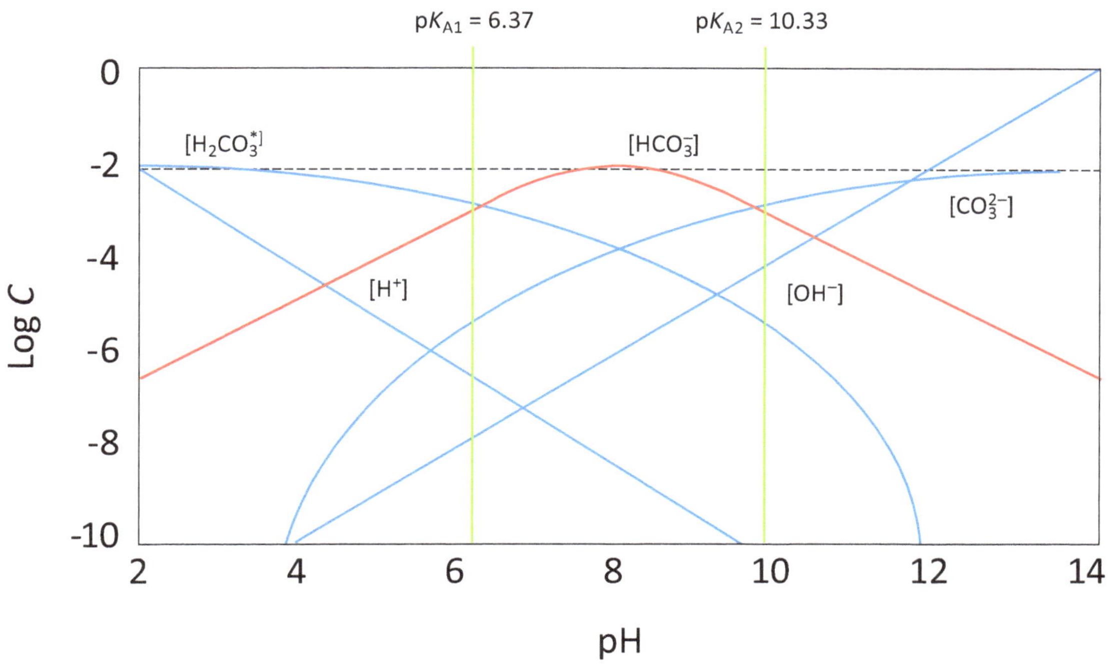

Figure 5-9 Alkalinity diagram

Source: Courtesy of Frederick Bloetscher

Figure 5-10 Conical accelator unit

for reasons that include cost savings and a lack of understanding, the lime softening system is not as effective as it should be. Many issues result:

- turbid water,
- biofilms,
- excessive sludge that is dilute,
- fouled filters,
- increase in filter grain size,
- filter plugging,
- cementation of filters,
- fouled ion-exchange resins, and
- excessive chlorine dose.

AWWA has a video that goes over the process but, given that lime softening is purely a groundwater issue, an example is worth pursuing. The lime/hardness reaction is a fast, floc-forming reaction. However, it slows as the pH drops.

Source: Courtesy of Frederick Bloetscher

Figure 5-11 Lime sludge

After the lime is added, the pH decreases from more than 10 to 9.5 because of settling of the metal hydroxides created by the lime reaction (lowering the pH). Hence, excess lime is always added per the example provided in Appendix A. Most utilities do not want the pH at 9.5. To reduce it to 8.6–9, the treated discharge from a lime or lime–soda ash process softener has carbon dioxide bubbled through it (recarbonation) to stabilize the effluent. Acid is also used to convert carbonates to bicarbonates and render the water stable. The pH should be above 8.6 to limit biofilm growth. Because pH is a secondary standard, the higher pH does not trigger any regulatory compliance issues.

Figures 5-10 to 5-13 show the accelator unit typically used for lime softening. The conical shape helps the lime sludge settle (Figure 5-10). The lime sludge is a yellow, sticky material with the consistency of cheesecake (Figure 5-11). The top of the unit includes the mixing zone in the center and the clarifier rings (Figure 5-12). Figure 5-13 shows a unit with no water in it.

A benefit of the lime softening process is that despite open basins, the lime–soda ash process can be used at ambient or elevated temperatures.

Source: Courtesy of Frederick Bloetscher

Figure 5-12 Top of accelator unit

Source: Courtesy of Frederick Bloetscher

Figure 5-13 Empty lime softening accelator unit

Iron and Manganese Treatment Methods

Ion exchange. Greensand continues to be a popular ion-exchange medium for the removal of iron and manganese. The process by which greensand removes these metals has been described as both ion exchange and adsorption. Generally, permanganate is added as a pretreatment to oxidize the metals and to regenerate the greensand. Chlorine is also used as a pretreatment, particularly if bacterial fouling of the filter medium is a problem.

Oxidation and filtration. If both the iron and manganese can be fully oxidized, the precipitates can be removed by filtration. However, manganese is more difficult to oxidize than iron, and complete oxidation requires adequate oxidation contact time. A typical treatment system includes increasing the pH before filtration to reduce the solubility of the metals.

Sequestering. Sequestering is a strategy for controlling iron and manganese by maintaining the metals in solution. Instead of removal, the metals are made complex by reacting with a sequestering agent to prevent their oxidation and precipitation. Phosphate or silicate compounds are often used for this purpose. The success of sequestering is site specific and generally limited to waters with a total concentration of iron and manganese below 1 mg/L.

Another option is diatomaceous earth (DE) filters. In this case, a porous material is "precoated" with a layer of DE at the beginning of a filter run. During the filter run, a "body feed" of DE is continuously fed with the influent water to build up the filter bed or cake. At the end of a filter run, the filter is backwashed, and the filter cake with the imbedded particles is sloughed off and discharged to waste.

Filtration

Filtration has traditionally been used to remove turbidity from surface water supplies. The following paragraphs discuss three common filtration methods often used in groundwater treatment: granular filtration, membrane filtration, and cake filters.

Granular filtration. Gravity sand filters are the most common means of filtration in the water treatment industry (see Figure 5-14). Filters remove solids from water using fine, porous media such as sand, anthracite coal, magnetite, garnet sand, and even coconut shells. Granular filtration is often preceded by lime softening. Groundwater systems may use pressure filters (see Figure 5-15) to achieve greater economic benefits over the open, gravity designs typically used for treatment of surface waters. All granular filters must be backwashed regularly to prevent the occurrence of "mudballs," which are congealed materials that stick together during backwash because they are too heavy to be removed.

Source: Courtesy of Frederick Bloetscher

Figure 5-14 Gravity sand filter

Source: Courtesy of Frederick Bloetscher

Figure 5-15 Pressure filter

Membrane filtration. Separation processes based on the use of selective membranes are becoming increasingly more important for the treatment of groundwater because of their ability to remove particles as small as metal ions without coagulation. Four major membrane processes are used for drinking-water treatment: microfiltration, ultrafiltration, nanofiltration, and reverse osmosis. Figure 5-16 illustrates the particle size ranges for these processes, and Figures 5-17 shows the process for a conceptual facility. Figure 5-18 shows an example installation. Pressurized raw water is fed to the membrane filters to produce a filtered water stream called the "permeate" and a wastewater stream called the "concentrate," which contains the concentrated contaminants rejected by the membrane.

Membrane filtration systems are used in desalination, softening, color removal, organics removal, and treatment of groundwater considered to be under the influence of surface water. Membranes are subject to fouling by iron, manganese, and microbiological activity. When present, these problems often require pretreatment, backwashing of the filter element, and periodic cleaning of the filter elements with a mild acid solution.

Cake filters. Cake filters, also known as precoat filters, deposit particles that become incorporated in the filter medium, which increases in depth. The best-known type is diatomaceous earth (DE) filters. DE is a naturally occurring material consisting of the microscopic remnants of the discarded frustules of diatoms.

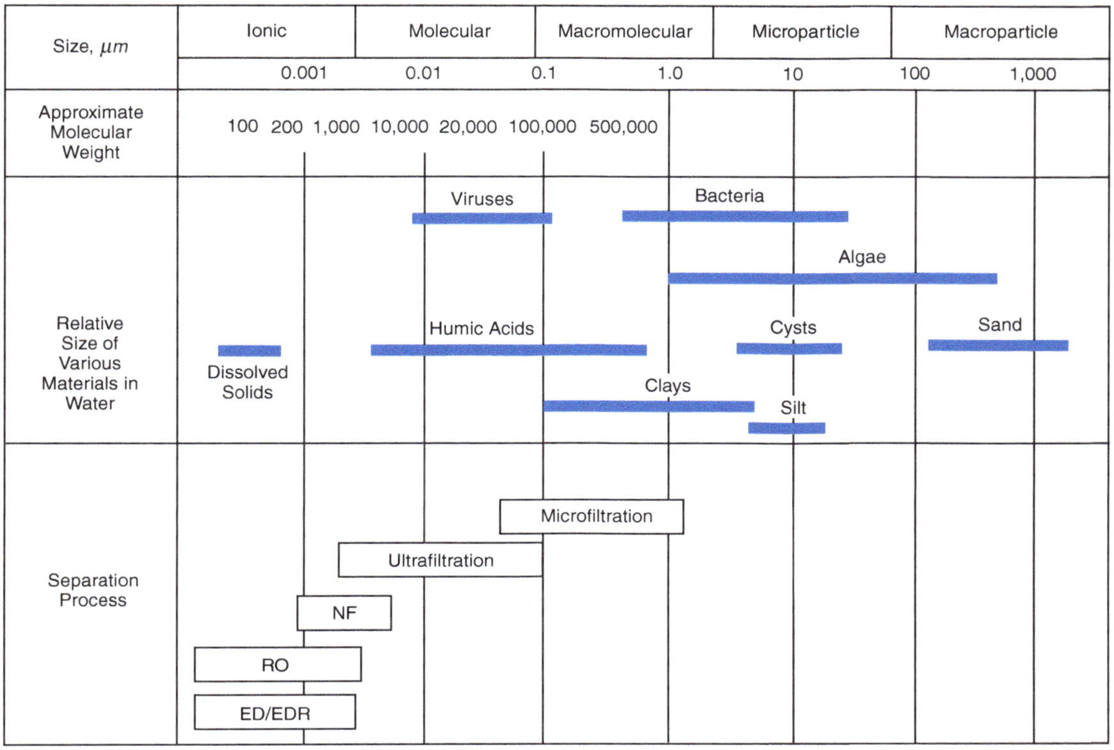

ED: electrodialysis, EDR: electrodialysis reversal, NF: nanofiltration, RO: reverse osmosis

Figure 5-16 Membrane and conventional process overview

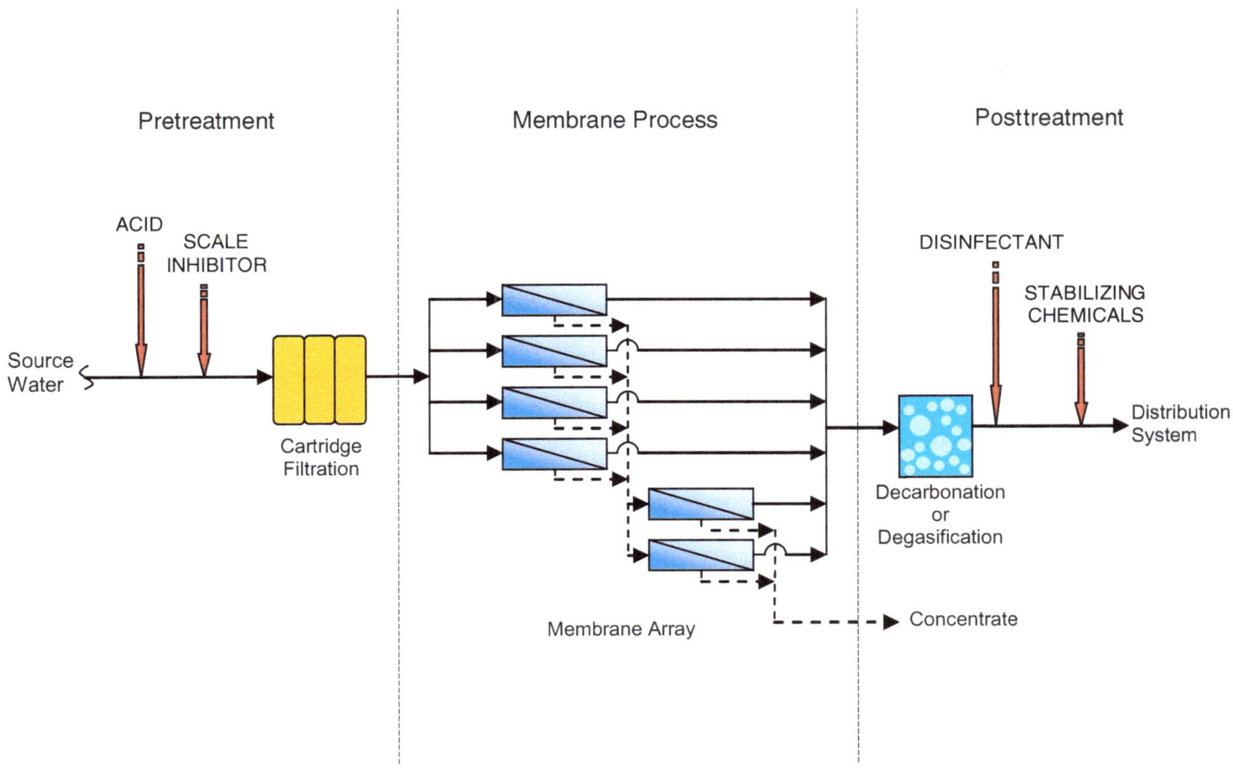

Figure 5-17 Typical reverse osmosis or nanofiltration membrane system

Source: Courtesy of Frederick Bloetscher

Figure 5-18 Membrane plant installation

Adsorption and Absorption

Adsorption is the collection of a substance onto the surface of another. This process is distinguished from absorption, which is the penetration of the substance into the solid. Both processes remove soluble contaminants; when occurring together, the combined process is referred to as "sorption." The most commonly used medium is activated carbon. Media have a limited adsorption and absorption capacity. When exhausted, contaminants break through and possibly cause desorption (leaching of contaminants). A major operating cost is the periodic replacement of the medium before its treatment capacity is reached.

Activated carbon. Removal of impurities by activated carbon involves both absorption and adsorption; however, in practice, the process is referred to as adsorption. Activated carbon is made from the carbonization or heating of various materials such as wood, sawdust, fruit pits and coconut shells, coal, and petroleum-based residues. The solids are carbonized and then activated using hot air or steam-producing pores, which increases the effective surface area per unit mass of carbon. Powdered activated carbon (PAC) and granular activated carbon (GAC) are used in drinking-water treatment.

GAC. GAC is used in fixed beds in either pressure or open, gravity filters. GAC is often used to treat taste, odor, and color problems and is considered by the USEPA to be the best available technology (BAT) for removal of many regulated VOCs and SOCs. GAC is also used for removal of radon-222 and disinfection byproducts. Potential problems include breakthrough, desorption of contaminants, the effects of backwashing on GAC loss, and bacterial growth on the filter medium. Spent GAC may be regenerated and reactivated. GAC may be part of the sand filters.

PAC. PAC differs from GAC by its smaller particle size and application. PAC is added as a dry powder or slurry before filtration. PAC is added only when needed, often to treat sporadic or seasonal taste and odor problems. PAC is rarely used for groundwater.

Note: GAC is suggested as one treatment option for short-chain PAS compounds, but the GAC cannot be recharged if PFAS are present.

Corrosion Control

Corrosion is the deterioration of metallic structures in contact with water, usually with loss of metal to solution. External pipeline corrosion is important; however, this discussion focuses on internal pipeline corrosion. Internal corrosion can result from metabolic (microbial) activity, chemical dissolution, or physical abrasion by excessive fluid velocities. Corrosion can affect both the structural capacity of the pipe and the quality of the water.

Corrosion typically causes rusting, pitting, and tuberculation of iron, copper, and lead water pipes, valves, and appurtenances. Metal is leached into the water. The severity of the problem depends on the chemical corrosivity of the water, biological activity, types of pipe materials, and other factors.

Water is considered chemically corrosive if

- pH, alkalinity, hardness, silicates, and phosphates are relatively low or
- dissolved oxygen, chlorine residual, TDS (or specific conductance), chlorides, sulfates, hydrogen sulfide, and carbon dioxide are relatively high.

Several indexes express the degree of corrosivity of water. The Langelier Saturation Index (LSI) is based on the effect of pH on the solubility of calcium carbonate. A positive LSI indicates a noncorrosive water (i.e., one that would precipitate a protection coating of calcium carbonate on pipes); a negative LSI indicates a corrosive water (i.e., one that would dissolve calcium carbonate). The calculation of the LSI is based on the alkalinity, calcium, TDS, pH, and temperature of the water.

Treatment for corrosion. Strategies for reducing corrosivity generally include increasing the pH, increasing the alkalinity, or adding a corrosion inhibitor. Common chemical additives for increasing the pH and alkalinity include lime, soda ash, sodium bicarbonate, caustic soda, and potassium hydroxide. Common corrosion inhibitors include silicates, orthophosphate, polyphosphate, and phosphate blends, which react with the pipe material to form a less soluble metal coating. This action is known as *passivation*. When adding silicates or phosphates, an adequate chlorine residual must be maintained because these compounds are nutrients that can stimulate microbiological activity and lead to taste and odor complaints.

Disinfection

Disinfection is defined as the destruction of *pathogenic* microorganisms (as opposed to sterilization, in which all living organisms are destroyed). Bacteria, viruses, protozoan, amoebic cysts, algae, and helminth (worms) are targeted organisms. Disinfection is most commonly achieved using chemical oxidizing agents such as chlorine, chlorine dioxide (ClO_2), chloramines, iodine, and ozone. Bacteria and viruses are the most common in groundwater—the others are issues only if the groundwater is under the influence of surface water. Other methods include UV radiation and maintenance of an elevated pH.

Disinfection effectiveness depends on the sensitivity of targeted microorganisms, disinfection concentration, contact time, and other water quality characteristics. Bacteria are the most sensitive to disinfection, followed by viruses, protozoan spores, and bacteria spores. Some enteric viruses that lack sensitive enzyme systems are very resistant.

Source: Courtesy of Frederick Bloetscher

Figure 5-19 Typical chlorine cylinder installation

Traditionally, the presence of coliform bacteria has been used as an indicator of viruses and other microorganisms. However, this practice is under review following outbreaks of some enteric viruses and the protozoan *Cryptosporidium* without the presence of coliforms.

Chlorine. Chlorine is the most widely used disinfectant because it is effective at low concentrations, is relatively inexpensive, and can form a residual. Chlorine can be applied as a gas or as hypochlorite (see Figures 5-19 and 5-20). Hypochlorite salts are available in dry (calcium hypochlorite) or liquid (sodium hypochlorite) form. Gas chlorine does not change, but hypochlorite does. The normal delivery is a 12.5 percent solution. Sunlight and time will degrade the hypochlorite to a 6 percent solution. As a result, many operators will blend it immediately to a 6 percent solution. Tankage is needed to accomplish this—some regulatory agencies require 30 days of storage.

Source: Courtesy of Frederick Bloetscher

Figure 5-20 Hypochlorite installation

When mixed with water, chlorine forms hypochlorous acid (HOCl) and hydrochloric acid (HCl). The HOCl further disassociates to yield the hypochlorite ion (OCl^-). Compared with OCl^-, HOCl is more effective. Formation of the hypochlorite ion from HOCl is pH dependent. At pH 7, 80 percent of the chlorine exists as HOCl; at pH 8, 80 percent of the chlorine exists as the less effective OCl^-. Effective chlorination requires careful attention to system pH.

Chlorine is a very reactive oxidizing agent and combines with ammonia, sulfites, metals, and organic material. Chlorine demand is the amount of chlorine that is used up in these extraneous reactions before it becomes free available chlorine for use as a disinfectant.

Chlorination reactions may produce byproducts, including THMs and organic halides in waters that contain humics or other natural organic precursors. If byproduct formation exceeds regulated limits, options include precursor removal or the use of alternative disinfectants.

Chloramines. Chloramines are compounds formed when chlorine reacts with ammonia, which may be naturally present or intentionally added to water. Customers should be informed because ammonia is toxic to fish (i.e., fish tanks). Chloramines are less effective than HOCl and OCl^- and less effective against viruses than bacteria. The benefits of chloramine disinfection are a lower generation of chlorination byproducts and greater residual stability in the distribution system. The pH of chloraminated waters should be kept above a pH of 8.6 to reduce the potential for biofouling.

Chlorine dioxide. Chlorine dioxide has not been widely used as a disinfectant. Like chloramines, chlorine dioxide is not as effective a disinfectant as chlorine. However, chlorine dioxide does not react with ammonia or nitrogenous compounds, nor does it react with carbon precursors to produce THMs. That is a benefit in certain situations. However, chlorine dioxide does produce two byproducts—chlorite (ClO_2^-) and chlorate (ClO_3^-)—

which are candidates for future regulation. Chlorine dioxide has also been associated with odor generation in some homes with new carpeting. Chlorine dioxide must be generated onsite—it cannot be stored. Because of operational issues and regulatory uncertainty, chlorine dioxide is not commonly used in the United States.

Ozone. Most bacteria, including coliforms, are highly susceptible to inactivation by ozone. Exceptions are the relatively resistant gram-positive *Bacillus* and *Mycobacterium*. Viruses are generally more resistant than bacteria. Ozone is considered particularly effective for *Giardia lamblia* and *Cryptosporidium* cysts, which are relatively resistant to chlorine. Ozone does not produce a disinfection residual in the distribution system as chlorine does, so chlorine must be added. *Escherichia coli* are so sensitive to ozone relative to other organisms that they are not a good indicator of the quality of water disinfected by ozone.

Ozone can oxidize large, organic macropollutants into smaller, more biodegradable compounds, producing an increased level of AOCs. AOCs can stimulate distribution system biological activity, including increased biofilm production. Ozonation produces bromate (a regulated substance) in waters with bromide ions. The ions react with natural organic matter to produce tribromomethane and bromoacetic acids.

Fluoridation

A fluoride concentration of up to 1 mg/L in drinking water is generally considered to reduce dental decay. New recommendations suggest 0.8 mg/L is appropriate. Some groundwaters contain naturally high levels of fluoride, and in some cases, fluoride must be reduced to acceptable levels by ion exchange with activated alumina, lime softening, or membranes.

Hydrofluosilicic acid (H_2SiF_6) is often used as an additive because it is purchased as a bulk liquid. Dry additives include sodium fluoride (NaF), calcium fluoride (CaF_2), ammonium silicofluoride (NH_4SiF_6), and sodium silicofluoride (Na_2SiF_6). The dry additives require a dry feeder (either gravimetric or volumetric) and saturation tank or must be mixed by hand. Operators should be aware that dosage is related to the available fluoride, not the dilution of the fluid, so appropriate mathematics must be used.

REFERENCES

American Water Works Association (AWWA). 1990 (4th ed.). *Water Quality & Treatment: A Handbook on Drinking Water*. Denver: AWWA.

Anderson, L. 1990. "Groundwater Quality, A Regional Survey of Groundwater Quality in the Metropolitan Water District Service Area." Report No. 991. Los Angeles, Calif.: Metropolitan Water District of Southern California.

Bloetscher, F., D.E. Meeroff, K. Conboy, C.H. Sham, R.E. Gergen, R. Gallant, J. Hart, A. Muniz, T. Shibata, M.E. Tuccillo, and J.D. Englehardt. 2024. "Assessing Relative Risks of Municipal Wastewater Disposal Options for Southeast Florida." *Risk Analysis* 2024; 1–21. https://doi.org/10.1111/risa.14301

Bloetscher, F. and J.D. Plummer. 2011. "Evaluating the Significance of Certain Pharmaceuticals and Emerging Pathogens in Raw Water Supplies." *Environmental Practice*. 13(3):198–215.

Bloetscher, F., D. Stambaugh, J. Hart, J. Cooper, K. Kennedy, L.S. Burack, A.P. Ruffini, A. Cicala, and S. Cimenello. 2011. "Evaluating Membrane Options for Aquifer Recharge in Southeast Florida." *IDA Journal of Desalination and Water Reuse*. 3(4):46–57.

Bloetscher, F., F. Pleitez, J. Hart, D. Stambaugh, J. Cooper, K. Kennedy, and L.S. Burack. 2014. "Comparing Contaminant Removal Costs for Aquifer Recharge with Wastewater with Water Supply Benefits." *Journal of the American Water Resources Association.* 50(2):324–333.

Buser, H.-R., T. Poiger, and M.D. Müller 1999. "Occurrence and Environmental Behavior of the Chiral Pharmaceutical Drug Ibuprofen in Surface Waters and in Wastewater." *Environmental Science & Technology.* 33(15):2529–2535.

Chapelle, F.H. 1993. *Groundwater Microbiology and Geochemistry.* New York: John Wiley and Sons.

Colburn, T., D. Dumanosui, and J.P. Myers. 1997. *Our Stolen Future.* New York: Plume.

Daughton, C.G. and T.A. Ternes. 1999. "Pharmaceuticals and Personal Care Products in the Environment: Agents of Subtle Change?" *Environmental Health Perspectives.* 107(Suppl. 6): 907–938.

Depledge, M.H. and Z. Billinghurst. 1999. "Ecological Significance of Endocrine Disruption in Marine Invertebrates." *Marine Pollution Bulletin.* 39(1–12):32–38.

Durfor, C.N. and E. Becker, 1964. "Public Water Supplies of the 100 Largest Cities in the United States, 1962." Geological Survey Water-Supply Paper 1812. Washington, D.C.: US Geological Survey.

Feachem, R., H. Garelick, and J. Slade. 1981. "Enteroviruses in the Environment." *Tropical Diseases Bulletin.* 28(3)185-230.

Guillette, L.J., T.S. Gross, G.R. Masson, J.M. Matter, H.F. Pierce, and A.R. Woodward. 1994. "Developmental Abnormalities of the Gonad and Abnormal Sex Hormone Concentrations in Juvenile Alligators from Contaminated and Control Lakes in Florida." *Environmental Health Perspectives.* 102(8):680-688.

Hart, J., D. Stambaugh, J. Cooper, L.S. Burack, F. Bloetscher, and S. Cimenello. 2011. "Membrane and Solute Combinations for ESOC Removal Prediction." FSAWWA Conf. Proc., Orlando, Fla.

Hirsch, R., T.A. Ternes, K. Haberer; and K.-L. Kratz. 1999. "Occurrence of Antibiotics in the Aquatic Environment." *The Science of the Total Environment.* 225(1-2):109-118.

Hurst, C.J. 1996. *Modeling Disease Transmission and Its Prevention by Disinfection.* Cambridge, UK: Cambridge University Press.

Lutz, I. and W. Kloas. 1999. "Amphibians as a Model to Study Endocrine Disruptors: I. Environmental Pollution and Estrogen Receptor Binding." *Science of the Total Environment.* 225(1–2):49–57.

Montgomery, J.H. 1996. *Groundwater Chemicals Desk Reference.* New York: Lewis Publishers.

Raloff, J. 2001. "Macho Water: Some River Pollution Spawns Body-Altering Steroids." *Science News.* 159(1):8–10.

Ternes, T.A. 1998. "Occurrence of Drugs in German Sewage Treatment Plants and Rivers." *Water Research*. 32(11):3245-3260.

National Toxicology Program. 1999 (9th ed.). "Report on Carcinogens." Research Triangle Park, N.C.: National Toxicology Program.

USEPA. 1990. *Pesticide Fact Handbook, Volumes 1 and 2*. Park Ridge, N.J.: Noyes Data Corp.

www.epa.gov/ngispgm3/iris/subst/0045.html. USEPA IRIS website for NDMA as of November 2000.

www.epa.gov/qbuebdeu/ged/public/c3830.hmtl.

www.iatp.org/erdc/hormonedisruptos082599.htm.

BIBLIOGRAPHY

(USEPA) US Environmental Protection Agency, Office of Ground Water and Drinking Water. 2009. *National Public Drinking Water Standards*. Washington, D.C.: USEPA.

Chapter 6

Conclusion

The fifth edition of M21 is a complete revision of the scope of the previous M21 editions. In this edition, the focus has been shifted toward an operations guide for those systems that operate with wells. More than 41,000 water systems rely on wells—most of them small systems. This edition has been written to provide the reader with a general understanding of the principles involved in operations. The reader is urged to look at the recently produced M73 (AWWA 2021) for information on groundwater protection, planning, and modeling for sustainability. M63 (AWWA 2015) looks at aquifer storage and recovery. AWWA's well drilling book (Bloetscher et al. 2022) is a useful source for well drilling. Future committee efforts will focus on webinars, workshops, and webcasts on a variety of groundwater topics that pick up from these manuals.

The M21 title was changed to *Groundwater Operations* to reflect this new focus as well. This manual is intended to meet the needs of operations staff and engineers with sufficient information on the operations and monitoring of their wells to help them with decision-making before major problems arise.

Previous editions of M21 also talked little about treatment and did not include pictures. It was fitting that this effort be expended since there is an increase in groundwater treatment occurring as water quality diminishes or new aquifers with more challenging problems are developed. Groundwater is affected by new rules, especially with arsenic, other metals, and pathogen removal. Sections on water quality and well maintenance were also expanded as these are areas that will require more focus (water quality and well maintenance may be related); our understanding of microbially influenced deterioration continues to grow; and the challenges the subsurface organisms create increase. It is in this vein that the new M21 was born.

Groundwater remains the source of water for most smaller utilities and is where the greatest demand is increasing. It is also where it is becoming apparent that sustainability is not achieved. Recordkeeping is a necessary part of the effort to understand the subsurface and to plan for changes that may be needed, much like the eastern Carolinas did in the 1990s. The challenge is greater in the dry West. Understanding operations is critical to this effort. The committee hopes utilities and operations personnel find it useful.

REFERENCES

American Water Works Association (AWWA). 2015 (1st ed.). M63, *Aquifer Storage and Recovery*. Denver: AWWA.

AWWA. 2021 (1st ed.). M73, *Groundwater Management*. Denver: AWWA.

Bloetscher, F., A. Muniz, and J. Largey. 2022 (2nd ed.). *Water Supply Wells: Siting, Drilling, and Construction*. Denver: AWWA.

Appendix A

Lime Softening Calculation

As noted in Chapter 5, lime softening works well when the pH and lime doses are correct. Too often, for reasons that include cost savings and lack of understanding, the lime softening system is not as effective as it should be. Many issues result in:

- turbid water,
- biofilms,
- excessive sludge that is dilute,
- fouled filters,
- increase in filter grain size,
- filter plugging,
- cementation of filters,
- fouled ion-exchange resins, and
- excessive chlorine dose.

AWWA has a video ("Lime Softening Techniques for Operators: AWWA Thought Leaders Series DVD") that goes over the process. However, given that lime softening is purely a groundwater issue, an example is worth pursuing.

Suppose the water quality is presented as follows:

- Carbon dioxide—44 mg/L
- Calcium hardness—200 mg/L
- Magnesium ion—30 mg/L
- Sodium ion—10 mg/L
- Potassium—9 mg/l

- Alkalinity—250 mg/L
- Sulfate ion—65 mg/l
- Chloride ion—30 mg/L
- pH = 7.6

The doses of lime and soda ash from this example are unclear. The concentrations need to be converted to milliequivalents. Milliequivalents are defined as the molecular weight (MW) divided by the valence. If the constituent is listed as "calcium carbonate," the equivalent is 50. Using milliequivalents, the table would be as follows:

- Carbon dioxide—44 mg/L / (22/2) = 2.0 meq/L
- Calcium hardness—200 mg/L / (100/2) = 4.0 meq/L
- Magnesium ion—30 mg/L / (24.3/2) = 2.5 meq/L
- Sodium ion—10 mg/L / (23/1) = 0.45 meq/L
- Potassium ion—9 mg/L /(39/1) = .25 meq/L
- Bicarbonate Alkalinity—250 mg/L / (100/2) = 5 meq/L
- Sulfate ion—65 mg/L / (96/2) = 1.35 meq/L
- Chloride ion—30 mg/L / (35.5/1) = 0.85 meq/L

A milliequivalent graph can be constructed (Figure A-1). From this graph, the following combinations can be noted:

Lime softening assumes a series of equations:

$CO_2 + Ca(OH)_2 > CaCO_3 + 2H_2O$

$Ca(HCO_3)_2 + Ca(OH)_2 > 2CaCO_3 + 2H_2O$

$Mg(HCO_3)_2 + Ca(OH)_2 > CaCO_3 + MgCO_3 + 2H_2O$, which leads to the next equation:

$MgCO_3 + Ca(OH)_2 > Mg(OH)_2 + CaCO_3$

$MgCl_2 + Ca(OH)_2 > Mg(OH)_2 + CaCl_2$

$MgSO_4 + Ca(OH)_2 > Mg(OH)_2 + CaSO_4$

$CaSO_4 + Na_2CO_3 > CaCO_3 + Na_2SO_4$

$CaCl_2 + Na_2CO_3 > CaCO_3 + 2NaCl$

Using these equations, the lime and soda ash milliequivalents are noted from the equations and Table A-1. The milliequivalents can then be converted to lime and soda ash doses as follows:

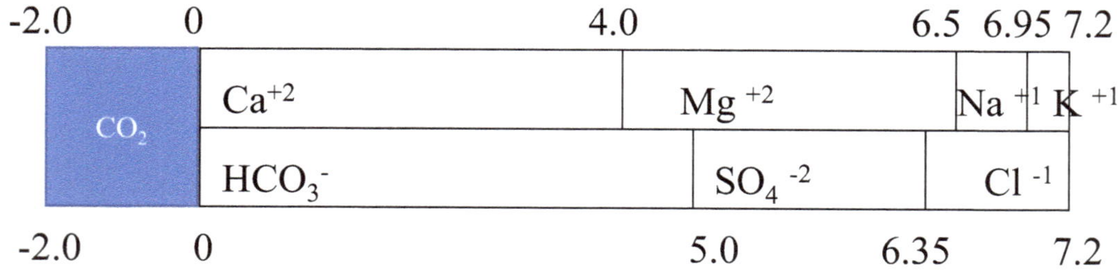

Source: Courtesy of Frederick Bloetscher

Figure A-1 Milliequivalent diagram

Table A-1 Milliequivalents of constituents

Constituent	Lime	Soda Ash
CO_2 – 2 meq/L	2	
$Ca(HCO_3)_2$ – 4 meq	4	
$Mg(HCO_3)_2$ – 1 meq	2	
$MgSO_4$ – 1.35 meq	1.35	1.35
$MgCl_2$ – 0.15 meq	0.15	0.15
Total (meq)	**9.5**	**1.5**

Because pH is such an important consideration, 35 mg/L of lime is added to the milliequivalent dose, since metal hydroxides will settle and reduce the pH. The lime alters the pH. Soda ash does not affect the pH but helps the calcium carbonate settle by adding needed alkalinity to the water. Care needs to be taken given the soda ash includes sodium. Often in areas with sodium in the water, soda ash is limited or eliminated.

Lime is CaO, which has an MW of 56 and an equivalent weight (Eq Wt) of 28 mg/meq, so:

9.5 meq × 28 mg/meq = 266 mg/L of lime + 35 mg/L = 301 mg/L lime

Always add an additional 35 mg/L of lime to keep the pH above 10.3.

Furthermore, soda ash is Na_2CO_3, which has an MW of 106 and an Eq Wt of 53 mg/meq, so:

1.5 meq × 53 mg/meq = 79.5 mg/L of soda ash

The resulting pH is around 9.5. Figure A-2 shows the resulting milliequivalent diagram.

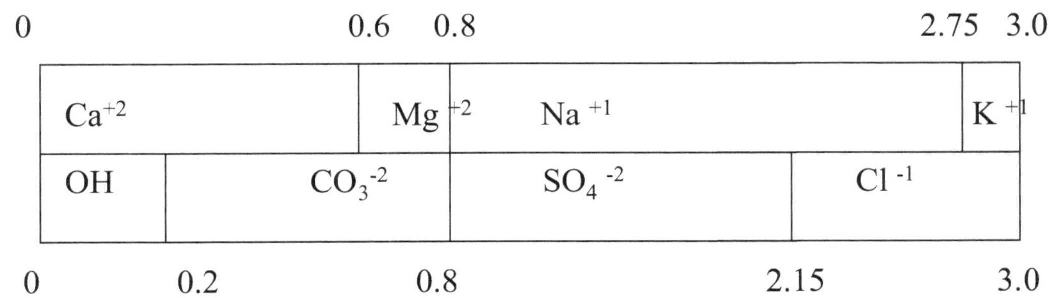

Source: Courtesy of Frederick Bloetscher

Figure A-2 Milliequivalent diagram after lime softening

This page intentionally blank.

Index

Note: *f* indicates figure; *t* indicates table

This page intentionally blank.

AWWA Manuals

M1, *Principles of Water Rates, Fees, and Charges*, #30001
M2, *Instrumentation and Control*, #30002
M3, *Safety Management for Water Utilities*, #30003
M4, *Water Fluoridation Principles and Practices*, #30004
M5, *Water Utility Management*, #30005
M6, *Water Meters—Selection, Installation, Testing, and Maintenance*, #30006
M7, *Problem Organisms in Water: Identification and Treatment*, #30007
M9, *Concrete Pressure Pipe*, #30009
M11, *Steel Pipe—A Guide for Design and Installation*, #30011
M12, *Simplified Procedures for Water Examination*, #30012
M14, *Backflow Prevention and Cross-Connection Control: Recommended Practices*, #30014
M17, *Fire Hydrants: Installation, Field Testing, and Maintenance*, #30017
M19, *Emergency Planning for Water and Wastewater Utilities*, #30019
M20, *Water Chlorination/Chloramination Practices and Principles*, #30020
M21, *Groundwater*, #30021
M22, *Sizing Water Service Lines and Meters*, #30022
M23, *PVC Pipe—Design and Installation*, #30023
M24, *Planning for the Distribution of Reclaimed Water*, #30024
M25, *Flexible-Membrane Covers and Linings for Potable-Water Reservoirs*, #30025
M27, *External Corrosion Control for Infrastructure Sustainability*, #30027
M28, *Rehabilitation of Water Mains*, #30028
M29, *Water Utility Capital Financing*, #30029
M30, *Precoat Filtration*, #30030
M31, *Distribution System Requirements for Fire Protection*, #30031
M32, *Computer Modeling of Water Distribution Systems*, #30032
M33, *Flowmeters in Water Supply*, #30033
M36, *Water Audits and Loss Control Programs*, #30036
M37, *Operational Control of Coagulation and Filtration Processes*, #30037
M38, *Electrodialysis and Electrodialysis Reversal*, #30038
M41, *Ductile-Iron Pipe and Fittings*, #30041
M42, *Steel Water-Storage Tanks*, #30042
M44, *Distribution Valves: Selection, Installation, Field Testing, and Maintenance*, #30044
M45, *Fiberglass Pipe Design*, #30045
M46, *Reverse Osmosis and Nanofiltration*, #30046
M47, *Capital Project Delivery*, #30047
M48, *Waterborne Pathogens*, #30048
M49, *Quarter-Turn Valves: Head Loss, Torque, and Cavitation Analysis*, #30049
M50, *Water Resources Planning*, #30050
M51, *Air Valves: Air-Release, Air/Vacuum and Combination*, #30051
M52, *Water Conservation Programs—A Planning Manual*, #30052
M53, *Microfiltration and Ultrafiltration Membranes for Drinking Water*, #30053
M54, *Developing Rates for Small Systems*, #30054
M55, *PE Pipe—Design and Installation*, #30055
M56, *Nitrification Prevention and Control in Drinking Water*, #30056
M57, *Algae: Source to Treatment*, #30057
M58, *Internal Corrosion Control in Water Distribution Systems*, #30058
M60, *Drought Preparedness and Response*, #30060
M61, *Desalination of Seawater*, #30061
M62, *Membrane Applications for Water Reuse*, #30062
M63, *Aquifer Storage and Recovery*, #30063
M65, *On-Site Generation of Hypochlorite*, #30065

M66, *Cylinder and Vane Actuators and Controls—Design and Installation,* #30066

M68, *Water Quality in Distribution Systems,* #30068

M69, *Inland Desalination and Concentrate Management,* #30069

M77, *Condition Assessment of Water Mains,* #30077

www.ingramcontent.com/pod-product-compliance
Ingram Content Group UK Ltd.
Pitfield, Milton Keynes, MK11 3LW, UK
UKHW052012041026
15590UKWH00008B/2286

9 781647 171490